Voice-Data-Video
Applications and Installation

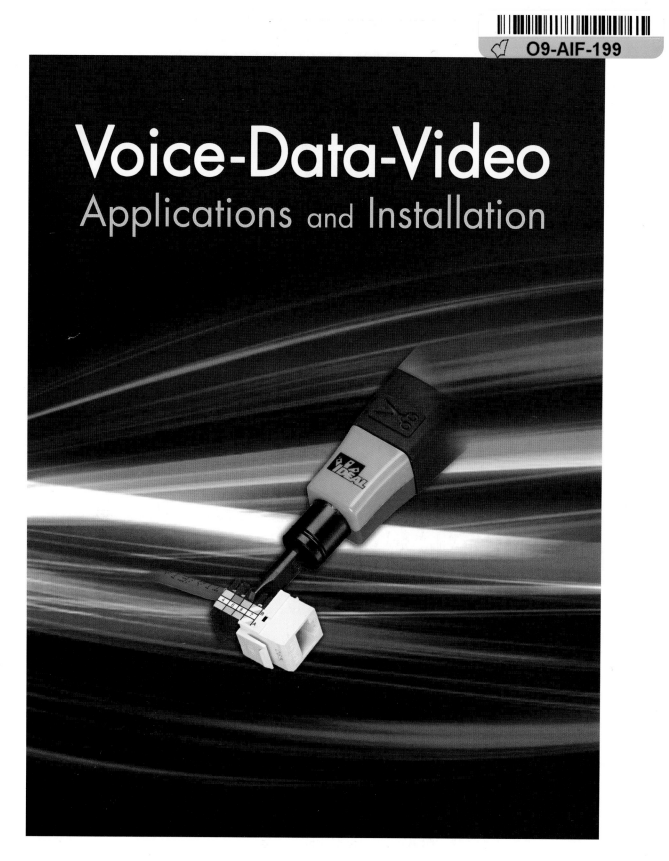

 AMERICAN TECHNICAL PUBLISHERS
Orland Park, Illinois

Matt Doell, P. E. RCDD
William J. Weindorf

Voice-Data-Video: Applications and Installation contains procedures commonly practiced in industry and the trade. Specific procedures vary with each task and must be performed by a qualified person. For maximum safety, always refer to specific manufacturer recommendations, insurance regulations, specific job site and plant procedures, applicable federal, state, and local regulations, and any authority having jurisdiction. The material contained herein is intended to be an educational resource for the user. American Technical Publishers, Inc. assumes no responsibility or liability in connection with this material or its use by any individual or organization.-

American Technical Publishers, Inc., Editorial Staff

Editor in Chief:
Jonathan F. Gosse
Vice President—Editorial:
Peter A. Zurlis
Assistant Production Manager:
Nicole D. Bigos
Technical Editor:
James T. Gresens
Supervising Copy Editor:
Catherine A. Mini
Copy Editor:
James R. Hein
Editorial Assistant:
Sara M. Patek

Cover Design:
Bethany J. Fisher
Art Supervisor:
Sarah E. Kaducak
Illustration/Layout:
Christopher S. Gaddie
Nick G. Doornbos
Nicholas W. Basham
Steven E. Gibbs
Digital Media Manager:
Adam T. Schuldt
Digital Resources:
Lauren M. Lenoir

1 2 3 4 5 6 7 8 9 – 17 – 9 8 7 6 5 4 3 2 1

Printed in the United States of America

ISBN 978-0-8269-1827-7

This book is printed on recycled paper.

Acknowledgments

The author and publisher are grateful for the photographs and technical information provided by the following companies and organizations:

Abaco Systems
Beast Cabling Systems
CSI Masterformat
Fluke Corporation
Fluke Networks
Greenlee Textron, Inc.
Honeywell International, Inc.
IDEAL Industries, Inc.
Klein Tools, Inc.
Leviton Manufacturing Co., Inc.
Linden Group, Inc.

Milwaukee Tool Corp.
NREL
Panduit Corp.
Salisbury
Salisbury by Honeywell
Square D/Schneider Electric
Shieldlab USA
The Stanley Works
Sonny's Enterprises, Inc.
Test Um, Inc.

Contents

Chapter 4

Copper Structured Cabling Systems 53

Low-Voltage Cable Evolution • Twisted-Pair Cables • Cable Categories • Color Codes • Connectors • Backbone Twisted-Pair Cables • Shielded-Pair Cables • Coaxial Cables • National Electrical Code® Ratings • ANSI/TIA/EIA 568 Requirements • Warranties and Packaging

Chapter 5

Fiber-Optic Cabling Systems 71

Fiber-Optic Systems vs Copper-Based Systems • Fiber-Optic System Advantages • Overall System Cost • Technical Complexity • Fiber-Optic Cable Construction • Light Modes • Multimode Fiber • Single-Mode Fiber • Fiber-Optic Performance Factors • Attenuation • Modal Dispersion • Chromatic Dispersion • Acceptance Angle • Numerical Aperture • NEC® Fiber-Optic Cable Designations • Fiber-Optic Cable Types • Tight vs Loose Cable • Ribbon Fiber-Optic Cables • Indoor Cables and Outdoor Cables • Cable Jacket and Sheath Colors • Fiber-Optic Cable Strand Counts • Fiber-Optic Connectors • Connector Performance • Connector Styles • Fiber-Optic Safety • Fiber-Optic Digital Transmissions • Optical Transmitters and Receivers • Cable Bend Radius • Light Intensity, Propagation, and Measurement • Fiber-Optic Cable Warranties

Chapter 6

VDV Prints 93

VDV System Prints • VDV Abbreviations, Symbols, and Legends • VDV Riser Diagrams • VDV Floor Plans • VDV Detail Drawings

Introduction

Voice-Data-Video: Applications and Installation presents a comprehensive overview of low-voltage cabling, devices, and circuitry used to transmit voice, data, and video. This new textbook is designed for electricians, students, and technicians who have a basic understanding of electricity and voice-data-video (VDV) applications.

Connectors, cables, circuit construction, operation, installation, termination, and troubleshooting are emphasized and supported with detailed illustrations. Various practical applications and procedures are presented throughout the book as they relate to copper and fiber-optic VDV systems. This book begins with electrical safety and an introduction to the VDV industry and continues with the presentation of the separate fundamentals of copper and fiber-optic systems. VDV prints and codes are covered as well as grounding and bonding of VDV systems. The most commonly used cables, connectors, and test instruments are depicted.

Chapter objectives at the beginning of each chapter provide learning goals related to the topics discussed. Detailed illustrations depict components, devices, systems, operations, and testing procedures. Chapter summaries at the end of each chapter provide a brief synopsis of the material covered. A review at the end of each chapter includes relevant questions, and an activity helps reinforce understanding of chapter material.

Digital resources include Quick Quizzes®, an illustrated glossary, flash cards, termination and testing procedures, a media library that consists of animations and videos, and access to ATP eResources to enhance learner comprehension.

Features

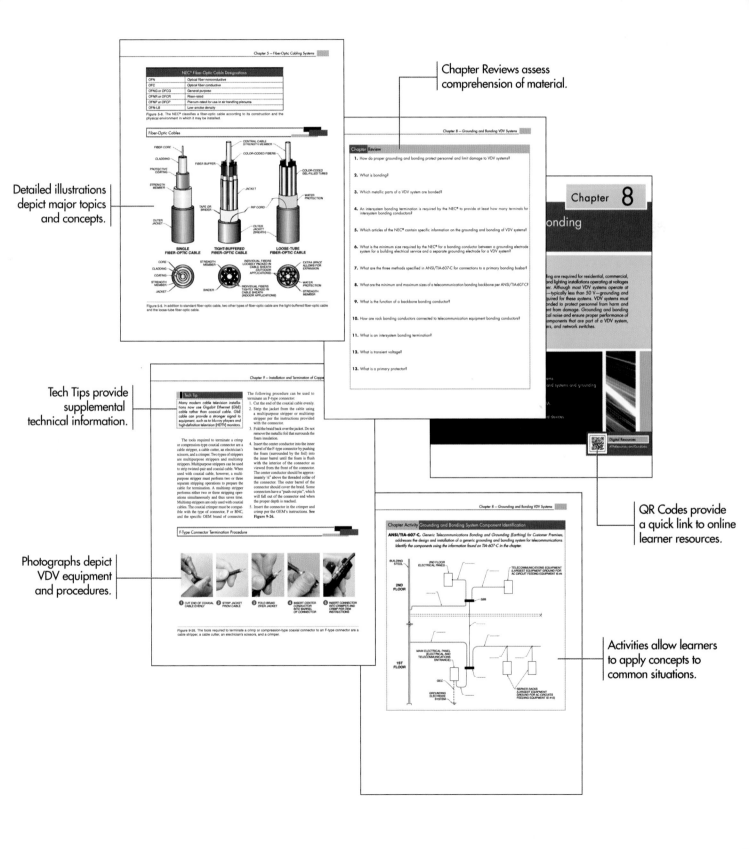

Chapter Reviews assess comprehension of material.

Detailed illustrations depict major topics and concepts.

Tech Tips provide supplemental technical information.

Photographs depict VDV equipment and procedures.

QR Codes provide a quick link to online learner resources.

Activities allow learners to apply concepts to common situations.

Learner Resources

Voice-Data-Video: Applications and Installation includes access to online Learner Resources that reinforce content and enhance learning. These online resources can be accessed using either of the following methods:

- Key ATPeResources.com/QuickLinks into a web browser and enter QuickLinks™ Access code 838502.
- Use a Quick Response (QR) reader app to scan the QR Code with a mobile device.

The Learner Resources include the following:

- **Quick Quizzes®** that provide interactive questions for each chapter, with embedded links to highlighted content within the textbook and to the Illustrated Glossary
- **Illustrated Glossary** that serves as a helpful reference to commonly used terms, with selected terms linked to textbook illustrations
- **Flash Cards** that provide a self-study/review of common terms and their definitions
- **Termination and Testing Procedures** that depict actual terminations of common VDV connectors and receptacles
- **Media Library** that consists of videos and animations that reinforce textbook content
- **ATPeResources.com**, which provides access to additional online resources that support continued learning

To obtain information on other related training material, visit the American Technical Publishers website at www.atplearning.com.

The Publisher

Chapter 1

Electrical and VDV Safety

Voice-data-video (VDV) technicians perform work at a wide variety of residential, commercial, and industrial job sites. Their work includes both new construction projects and retrofits to existing premises. VDV technicians must follow Occupational Safety and Health Administration (OSHA) safety rules, company safety policies, and job-site-specific safety rules. While each job location is unique, many safety considerations are universal. These include wearing personal protective equipment, using proper lifting procedures, practicing ladder safety, using safety barriers, referring to safety data sheets, and practicing electrostatic discharge precautions.

OBJECTIVES

- Describe how to use and maintain personal protective equipment.
- Explain the proper use of ladders and safety barriers.
- Explain how to use safety data sheets.
- Describe proper electrostatic discharge precautions.

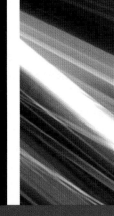

Digital Resources

ATPeResources.com/QuickLinks
Access Code 838502

PERSONAL PROTECTIVE EQUIPMENT

Personal protective equipment (PPE) is clothing, glasses, gloves, hard hats, respirators, or other safety devices designed to protect workers against safety hazards in the work area and from injury. PPE appropriate to the task must be worn when installing any VDV system. All PPE must meet OSHA Standard Part 1910 Subpart I–*Personal Protective Equipment* (1910.132 through 1910.138), applicable ANSI standards, and other safety mandates. **See Figure 1-1.** PPE involves high-visibility clothing, head protection, eye protection, ear protection, hand and foot protection, and knee protection.

High-Visibility Clothing

High-visibility clothing is a class of clothing made with extremely bright, color-enhanced fabric and includes shirts, pants, vests, and jackets. Typical colors include highly visible shades of orange, lime, or yellow. Some types of high-visibility clothing have vertical and horizontal stripes of reflective material for enhanced visibility.

Poor lighting on job sites can contribute to accidents or unintentional collisions between personnel or between personnel and equipment. VDV technicians wear high-visibility clothing so they can be seen readily by other tradesworkers, mobile work platform operators (such as scissor lift or articulated lift operators), crane operators, and heavy equipment operators.

Head Protection

Head protection involves wearing a protective helmet. A *protective helmet* is a rigid hat made from plastic that is used in the workplace to prevent head injury from impact or penetration by falling and flying objects. These helmets also protect technicians by preventing the wearer's head from coming into contact with electrical conductors. Protective helmet shells are durable and lightweight. A shock-absorbing lining, consisting of crown straps and a headband, keeps the shell slightly over the head to provide ventilation.

Protective helmet standards indicate the level of protection a helmet provides against specific hazardous conditions. Standards for protective helmets are specified in ANSI/ISEA Z89.1, *American National Standard for Industrial Head Protection*. (OSHA regulations also require technicians to cover and protect long hair in order to prevent it from getting caught in moving machine parts, such as belts and chains.)

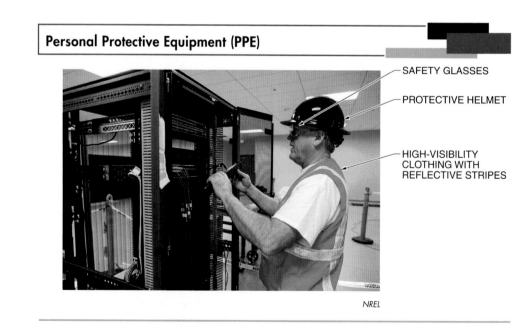

Personal Protective Equipment (PPE)

— SAFETY GLASSES

— PROTECTIVE HELMET

— HIGH-VISIBILITY CLOTHING WITH REFLECTIVE STRIPES

NREL

Figure 1-1. PPE must always be worn when working on job sites.

Protective helmets are divided into classes based on their uses. Helmets that provide electrical protection are classified as Class C, Class G, or Class E. Class C helmets are manufactured with lighter materials yet provide adequate impact protection. Class G helmets are used for general service applications and provide limited voltage protection. Class E helmets are used for utility applications and provide high-voltage protection. **See Figure 1-2.**

Protective Helmets

Class	Use
C	Special service, no voltage protection
E	Utility service, high voltage protection
G	General service, limited voltage protection

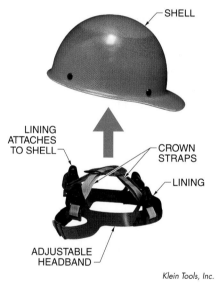

Klein Tools, Inc.

Figure 1-2. Protective helmet standards indicate the level of protection a helmet provides against specific hazardous conditions.

Eye Protection

Eye protection must be worn to prevent eye or face injuries caused by flying particles or debris. Devices in this category must comply with OSHA 29 CFR 1910.133, *Eye and Face Protection*. Eye-protection standards are specified in ANSI Z87.1, *Occupational and Educational Eye and Face Protection*. Eye-protection devices include safety glasses, face shields, and goggles. **See Figure 1-3.**

Eye Protection

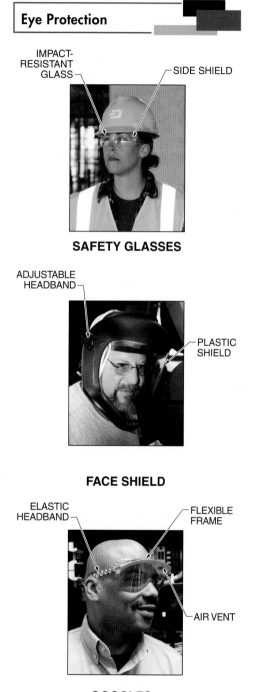

Figure 1-3. Eye protection must be worn to prevent eye or face injuries caused by flying particles or debris.

Earmuffs are a tight-fitting device worn over the ears for protection from excessive noise levels.

Safety glasses are an eye-protection device with reinforced frames, side shields, and lenses made of impact-resistant glass or plastic. The reinforced frames are also made of plastic and are designed to keep the lenses secured in the frame if an impact occurs. They also reduce the hazard of shock when working with electrical equipment.

A *face shield* is an eye-and-face protection device that covers the wearer's entire face with a plastic shield. It is used to protect the face from flying objects.

Goggles are an eye-protection device with a flexible frame that is secured to the wearer's face with an elastic headband. They protect against small flying particles and splashing liquids. Goggles fit snugly against the face to seal the areas around the eyes and may be used over prescription glasses.

Safety glasses, face shields, and goggles must be properly maintained so they provide protection and do not obstruct the wearer's vision. Lens cleaners are available that clean without risk of lens damage. Pitted, crazed, or scratched lenses reduce vision and may fail on impact.

Hearing Protection

Hearing protection is provided by devices worn to limit noise entering the ear such as earplugs and earmuffs. An *earplug* is a hearing ear-protection device made of moldable synthetic rubber or plastic foam that is inserted into the ear canal. *Earmuffs* are a hearing protection device worn over the ears. The tight seal formed over the ears by earmuffs is required for proper protection.

Power tools and equipment can produce excessive noise levels. Technicians subjected to excessive noise levels may develop hearing loss over a period of time. The severity of hearing loss depends on the intensity and duration of exposure. Noise intensity is expressed in decibels. A *decibel (dB)* is a unit of measure used to express the relative intensity of sound. **See Figure 1-4.**

Hearing-protection devices must be worn to prevent hearing loss due to excessive noise levels. These devices are assigned a noise reduction rating (NRR) based on the degree to which they reduce noise. For example, a device with an NRR of 27 reduces noise levels by 27 dB when tested at the original equipment manufacturer's (OEM) manufacturing facility. To determine the approximate noise reduction provided by a device in the field, 7 dB is subtracted from the NRR. For example, a device with an NRR of 27 provides a noise reduction of approximately 20 dB in the field.

Hand and Foot Protection

Hand protection is provided by safety gloves that are worn to prevent hand injuries caused by abrasions or cuts. Safety gloves are made entirely of animal skins (leather), entirely of synthetic materials, or a combination of both leather and synthetic materials.

The appropriate type of glove is determined by the degrees of protection and manual dexterity required. Safety gloves made from leather are typically used when a high degree of protection and a moderate degree of manual dexterity are required. Safety gloves made entirely from synthetic materials or from a combination of both leather and synthetic materials are normally used when a high degree of manual dexterity and a moderate degree of protection are required. **See Figure 1-5.**

Sound Levels			
Sound Intensity*	Description	Examples	Maximum Exposure Time
140	Deafening	Jet airplane taking off, air raid siren, locomotive horn	—
130	Pain threshold		2 min
120	Feeling threshold		7 min
110	Uncomfortable		30 min
100	Very loud	Chain saw	2 hr
90	Noisy	Shouting, auto horn	4 hr
85	Noisy	Snow thrower, diesel truck engine	8 hr
80	Moderately loud	Vacuum cleaner	25.5 hr
70	Loud	Telephone ringing, loud talking	—
60	Moderate	Normal conversation	—
50	Quiet	Hair dryer	—
40	Moderately quiet	Refrigerator running	—
30	Very quiet	Quiet conversation, broadcast studio	—
20	Faint	Whispering	—
10	Barely audible	Rustling leaves, soundproof room, human breathing	—
0	Hearing threshold	Intolerably quiet	—

* in dB

Figure 1-4. A decibel (dB) is a unit of measure used to express the relative intensity of sound.

Hand and Foot Protection

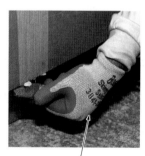

LEATHER/ SYNTHETIC

LEATHER *Salisbury by Honeywell*

SYNTHETIC

SAFETY GLOVES

SAFETY SHOES

Figure 1-5. Safety gloves and safety shoes are common types of PPE worn by VDV technicians.

Foot protection is provided by safety shoes that are worn to prevent foot injuries typically caused by objects falling a distance of less than 4′ and having an average weight of less than 65 lb. Safety shoes with reinforced steel toes provide protection from injuries caused by compression impact from a heavy object accidentally dropped on the foot or feet. Safety shoes must comply with ANSI Z41, *Personal Protection–Protective Footwear*.

Knee Protection

A *knee pad* is a rubber, leather, or plastic pad strapped onto the knee to provide protection and comfort as well as reduce fatigue. Knee pads are worn by technicians who spend considerable time working on their knees or who work in areas with limited space and must kneel for proper access. Buckle straps or hook-and-loop closures secure knee pads in position. **See Figure 1-6.**

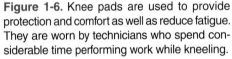

Knee Protection

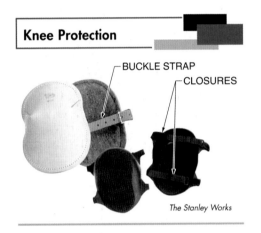

BUCKLE STRAP

CLOSURES

The Stanley Works

Figure 1-6. Knee pads are used to provide protection and comfort as well as reduce fatigue. They are worn by technicians who spend considerable time performing work while kneeling.

PROPER LIFTING PROCEDURES

A back injury is one of the most common injuries that results in lost time in the workplace. Back injuries are often the result of improper lifting procedures, but they can be prevented through proper planning. Assistance should be sought when moving heavy objects. When lifting heavy objects from the floor or ground level, apply the following procedure:

1. Ensure that the path of travel is clear of obstacles and free of hazards.
2. Bend the knees and grasp the object firmly.
3. Lift the object, straightening the legs and keeping the back as straight as possible.
4. Move forward after the entire body is in a vertical position.
5. Hold the load steady and close to the body while moving. Avoid twisting the body. **See Figure 1-7.**

Proper Lifting

1 ENSURE PATH OF TRAVEL IS CLEAR OF OBSTACLES AND FREE OF HAZARDS

2 BEND KNEES AND GRASP OBJECT FIRMLY

KEEP BACK STRAIGHT

3 LIFT OBJECT BY STRAIGHTENING LEGS

4 MOVE FORWARD AFTER ENTIRE BODY IS IN VERTICAL POSITION

5 HOLD LOAD STEADY AND CLOSE TO BODY WHILE MOVING

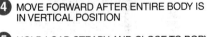

Figure 1-7. Lifting an object properly reduces the possibility of a back injury.

A long or bulky object may not be heavy, but the weight may not be distributed evenly. Therefore, such objects should be carried by two or more people.

LADDERS

A *ladder* is a device consisting of two side rails joined at intervals by steps or rungs. VDV technicians use ladders to perform tasks in work areas that are inaccessible from ground level. The most common type of ladder used by VDV technicians is the stepladder. A *stepladder,* or A-frame ladder, is a folding ladder that stands without support. Stepladders range in size from 2′ to 20′. **See Figure 1-8.** Only stepladders made of nonconductive material, such as wood or fiberglass, are used on job sites due to the possibility of accidental contact with electrical conductors. When a ladder is carried on the shoulder by one person,

it should be transported with the front end pointing downward to minimize the possibility of injury to others when walking around corners or through doorways. **See Figure 1-9.**

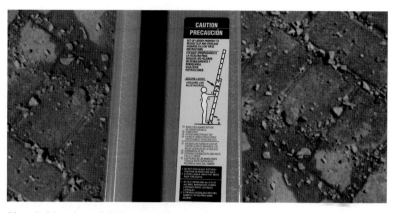

Most ladders have labels containing pertinent safety information affixed directly to one of the ladder's siderail(s).

Stepladders

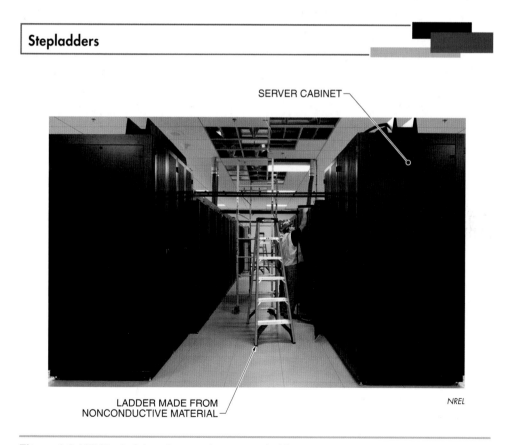

SERVER CABINET

LADDER MADE FROM NONCONDUCTIVE MATERIAL

NREL

Figure 1-8. VDV technicians frequently use stepladders.

Carrying Ladders

FRONT END POINTING DOWNWARD

Figure 1-9. When carried on the shoulder by one person, a ladder should be transported with the front end facing downward.

The ANSI A14 series of standards covers different types of ladders. Ladder duty ratings are described in the standards. A *ladder duty rating* is a rating that indicates the total weight, or the combined weight of personnel and tools, that a ladder is designed to support under normal use. The four ladder duty ratings are the following:

- Type IAA–extra heavy-duty, industrial, 375 lb capacity
- Type I–heavy-duty, industrial, 250 lb capacity
- Type II–medium-duty, commercial, 225 lb capacity
- Type III–light-duty, household, 200 lb capacity

Accidents involving ladders account for a significant amount of lost time in the workplace. It is extremely important to always follow all safety rules, regulations, and the ladder manufacturer's instructions when using a ladder. These include but are not limited to the following:

- Always inspect a ladder for broken parts or damage prior to use. Never use a defective ladder. If defects are found, tag the ladder as defective and remove it from service immediately.
- Only set a ladder on a firm, level surface.
- Fully open a stepladder.
- Only one person is allowed on a ladder at a time, unless it is rated as a two-man ladder.
- Do not overreach. Always keep the body centered between the side rails.
- Practice the three-point climbing method—with two hands and one foot or one hand and two feet in constant contact with the ladder—when ascending or descending.
- Do not stand on or above the top two steps of a stepladder.
- Always face a ladder when climbing or working.

SAFETY BARRIERS

A *safety barrier* is set of objects such as cones, line markings (delineators) or posts used to mark a potentially hazardous work area. Safety barriers are typically made of PVC, are orange in color, and may have reflective stripes. They are used in new construction projects as well as retrofit projects.

Caution tape and safety barriers are used to prevent personnel from entering work areas where potential hazards exist. For example, caution tape and safety barriers are often used to prevent personnel from entering an area where the floor tiles of a raised computer floor have been removed or where tradesworkers are working overhead. **See Figure 1-10.**

SAFETY DATA SHEETS

A *safety data sheet (SDS)* is a document that provides detailed information describing a chemical, instructions for its safe use, its potential hazards, proper disposal and appropriate first-aid measures. All chemical products are required to have an SDS, including aerosol spray cleaners, wire-pulling lubricants, and antioxidant compounds. The original equipment manufacturers (OEMs) of chemical products are required to furnish SDSs. OSHA requires SDSs for all chemical products used on a job site to be available and accessible to employees at the site.

Tech Tip

An SDS includes information on a chemical such as its properties; the physical, health, and environmental hazards it poses; the protective measures that can be taken if the chemical is involved in an emergency; and safety precautions for handling, storing, and transporting the chemical. Information contained in an SDS must be in English.

Safety Barriers

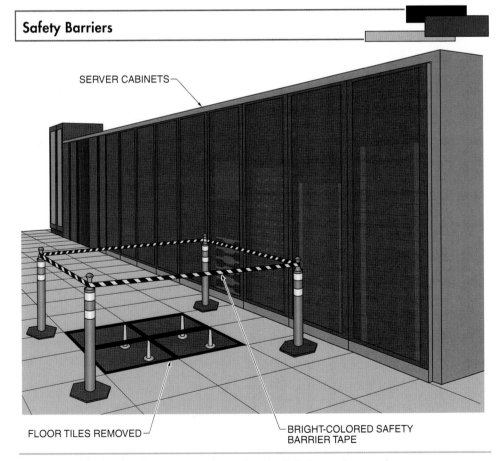

SERVER CABINETS

FLOOR TILES REMOVED

BRIGHT-COLORED SAFETY BARRIER TAPE

Figure 1-10. Safety barriers are used to mark potentially hazardous work areas.

SDSs are designed to inform and instruct employees in the safe handling of chemical products. In the event of an accident involving a chemical product, the SDS for that product should be consulted. **See Figure 1-11.** The length of an SDS, or the number of pages, varies with the hazard level of the product. Typically, the greater the hazard level, the greater the number of printed pages in an SDS. Information common to SDSs includes the following:

- Manufacturer's name, address, and phone number
- Product (trade) name
- Information on composition materials
- Reactivity data
- First aid procedures
- Firefighting procedures
- Accidental release procedures
- Handling and storage procedures
- Proper disposal procedures

ELECTROSTATIC DISCHARGE PRECAUTIONS

Static electricity is an electrical charge at rest. When materials rub together, the friction between them often creates a buildup of static electricity, or a negative charge, in one of the materials. For example, when a person walks across a rug, the person's shoes rub against the rug. This causes electricity to flow from the carpet to the person, and the person becomes negatively charged.

Safety Data Sheets (SDSs)

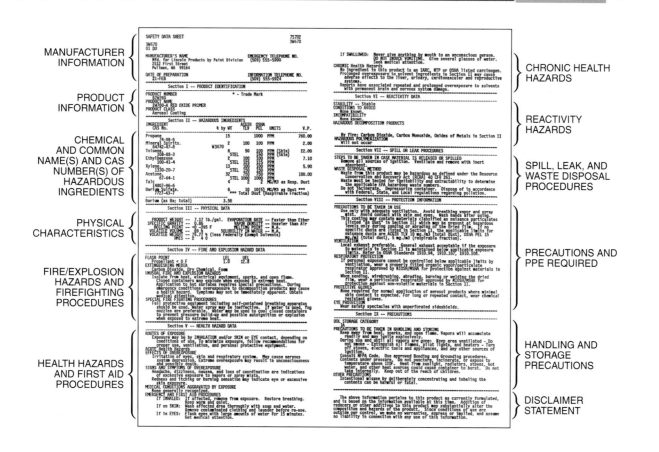

Figure 1-11. SDSs provide detailed information from manufacturers on the hazards and safe handling of chemical products.

When something that is negatively charged, such as a person, comes into contact with something that is positively charged, such as a grounded object, all excess electricity suddenly flows to the positively charged object. This is an electrostatic discharge. An *electrostatic discharge (ESD)* is a sudden movement of electricity between two objects. ESDs can be 35,000 V or more, but people normally do not feel them at voltages below 3000 V. **See Figure 1-12.**

Solid-state devices and circuits can be damaged or destroyed by a 10 V electrostatic discharge. Technicians should wear an antistatic wrist strap or other type of grounding device to avoid damage to solid-state devices and circuits from ESDs. The safety rules that are effective in preventing damage due to ESDs include the following:

- Wear an antistatic wrist strap to prevent static electricity buildup.
- Always keep a work area clean and clear of unnecessary materials, especially common plastics.
- Never handle electronic devices by their leads.
- Test grounding devices daily to verify that they have not become loose or intermittent.
- Never work on an ESD-sensitive object without a properly working grounding device.
- Always handle printed circuit boards, or PC boards, by their outside corners.
- Always transport PC boards in antistatic trays or antistatic bags.
- Never stack PC boards.
- When transporting single electronic devices, always keep them sealed in conductive static shielding.

Electrostatic Discharge

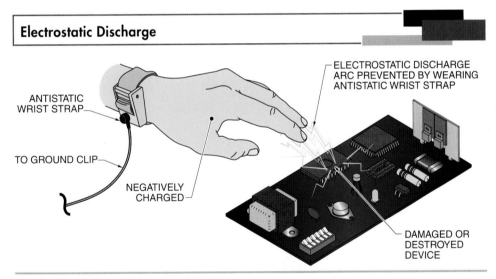

Figure 1-12. An electrostatic discharge produces an arc that can damage sensitive electronics.

Summary

To prevent damage to personnel and equipment, VDV technicians must always follow proper safety practices on the job. For protection, the use of high-visibility clothing to make oneself more visible is required. Technicians should always follow proper practices and wear proper PPE to protect the head, eyes, ears, hands, feet, and knees. Proper lifting and carrying procedures should be followed to prevent back injury. Additionally, proper use of ladders and safety barriers is required to prevent falling and collision hazards. SDSs must be referenced when handling various substances and materials on the job. Finally, electrostatic discharge precautions must be practiced to prevent damage from ESDs.

Chapter Review

1. List the seven items that comprise PPE.

2. What ANSI class of helmets provides no voltage protection?

3. List the three types of OSHA-approved eye-protection devices.

4. List the proper procedure for lifting and transporting a heavy object.

5. What is the ladder duty rating of a Type I ladder?

6. What should a VDV technician always do before using a ladder?

7. What is commonly used with safety barriers to prevent personnel from entering an area?

8. What is an SDS?

9. List three chemical products commonly used by VDV technicians that are required to have safety data sheets.

10. How should PC boards be transported in order to prevent damage from electrostatic discharge?

Chapter Activity Reading Safety Data Sheets (SDSs)

OSHA requires SDSs to be available and accessible to employees working on a job site for all chemical products used on the site. VDV technicians must know how to read an SDS for important safety information.

Note: Safety data sheets were previously referred to as material safety data sheets (MSDSs). However, not all manufacturers have updated their literature to reflect this change.

Refer to the material safety data sheet and answer the following questions.

Document # 6810-004

HMIS Rating

Flammability 0
Health 1
Reactivity 0

MATERIAL SAFETY DATA SHEET
GENERAL INFORMATION

PRODUCT NAME OR NUMBER (as it appears on label)
ClearGlide™ Wire Pulling Lubricant

CATALOG NUMBER
All "31" Series

MANUFACTURER'S NAME
IDEAL INDUSTRIES, INC.

EMERGENCY TELEPHONE NO.
(815) 895-5181

ADDRESS (Number, Street, City, State, Zip Code)
Becker Place, Sycamore, IL 60178

HAZARDOUS MATERIAL DESCRIPTION, PROPER SHIPPING NAME, HAZARD CLASS, HAZARD CLASS, HAZARD ID NO. (49 CFR 172.101)
None

CHEMICAL DESCRIPTION
Polymer-based Mixture

FORMULA
Proprietary

SECTION I - INGREDIENTS

CAS REGISTRY NO.	%W	CHEMICAL NAME(S)*	Listed as a carcinogen in NTP, I ARC or OSHA 1910(z) (specify)
7732-18-5	<98	Water	No
6440-58-0	<1	DMDM Hydantoin	No
9038-95-3	<5	Oxirane, Methyl-, Polymer With Oxirane, Monobutyl Ether	No
9003-11-6	<1	Amino-Methyl-Propanol	No
25322-68-3	<1	Polyethylene Glycol	No
9003-01-4	<1	Carbomer Thickener	No
68037-64-9	<1	Silicone Glycol Blend	No

SECTION II - PHYSICAL DATA

BOILING POINT	SPECIFIC GRAVITY (H₂O=1)	PERCENT VOLATILE BY VOLUME (%)
212°F °C	1.09	<98

SOLUBILITY IN WATER: Infinite
pH = 7.0 - 8.0
PERCENT SOLID BY WEIGHT (%): ~5
IS MATERIAL: LIQUID SOLID **GEL** GAS PASTE POWDER

APPEARANCE AND ODOR: Clear, colorless gel, slight odor

SECTION III - FIRE AND EXPLOSION HAZARD DATA

FLASH POINT None   method used C.O.C

FLAMMABLE LIMITS	LEL	UEL
	None	None

EXTINGUISHING MEDIA: Use extinguishing media suitable for surrounding materials.

SPECIAL FIRE FIGHTING PROCEDURES: None

UNUSUAL FIRE AND EXPLOSION HAZARDS: None

* None of the chemical raw materials contained in this formulation are considered hazardous under the Federal Hazards Communication Standard 29 C. F. R 1910.1200

form #5100-048-01, rev. date 3/10/06

IDEAL Industries, Inc.

Chapter Activity　Reading Safety Data Sheets (SDSs)

SECTION IV - HEALTH HAZARD INFORMATION

EFFECTS OF OVEREXPOSURE - Conditions to Avoid

None normally expected. Upon prolonged contact, may cause temporary eye discomfort.

THRESHOLD LIMIT VALUE

N.E.

PRIMARY ROUTES OF ENTRY Inhalation ☐ Skin Contact ☒ Other (specify)

EMERGENCY FIRST AID PROCEDURES

SKIN CONTACT:　Wash with soap and water for 15 minutes.

EYE CONTACT:

Flush with water for 15 minutes. Inhalation - Move to fresh air.

INGESTION:

Administer water or milk. Consult physician or local poison control center

SECTION V - REACTIVITY DATA

STABILITY	UNSTABLE		CONDITIONS TO AVOID
	STABLE	X	Avoid prolonged storage at temperatures exceeding 190 F.

INCOMPATIBILITY (materials to avoid)

Avoid strong oxidizers and nitrites

HAZARDOUS DECOMPOSTION PRODUCTS:

Oxides of carbon, nitrogen and silicone

HAZARDOUS POLYMERIZATION	MAY OCCUR		CONDITIONS TO AVOID
	WILL NOT OCCUR	X	None

SECTION VI - SPILL AND LEAK PROCEDURES

STEPS TO BE TAKEN IF MATERIAL IS RELEASED OR SPILLED

Wipe up, shovel or vacuum spilled material. Clean up spills immediately as they can be dangerously slippery.

WASTE DISPOSAL METHOD

Comply with Federal, state or local regulations for solid landfill

CERCLA (Superfund) REPORTABLE QUANTITY (in lbs)

N/A

RCRA HAZARDOUS WASTE NO. (40CFR 261.33)

N/A

VOLATILE ORGANIC COMPOUND (VOC) (as packaged, minus water)

17.4 gms / ltr

³ Theoretical _____ lb/gal   N/A   ³ Analytical _____ lb/gal   N/A

SECTION VII - PERSONAL PROTECTION INFORMATION

RESPIRATORY PROTECTION (specify type)

None normally required

VENTILATION	LOCAL EXHAUST (Specify Rate) None	SPECIAL None
	MECHANICAL (General) (Specify Rate) Recommended in closed areas.	OTHER None

PROTECTIVE GLOVES (specify type)　　EYE PROTECTION (specify type)

None normally required.　　　Safety glasses or splash goggles.

OTHER PROTECTIVE EQUIPMENT

Eye fountain in work area is recommended.

SECTION VIII - SPECIAL PRECAUTIONS

PRECAUTIONS TO BE TAKEN IN HANDLING AND STORING

Store at temperatures between 40 - 180 F. Avoid freezing

OTHER PRECAUTIONS

Keep away from children, infants and pets.

SECTION IX - ADDITIONAL INFORMATION

N/A = Not Applicable, N.E. = None Established

THIS MATERIAL SAFETY DATA SHEET PREPARED BY:

NAME	Darryl Docter	SIGNATURE
TITLE	Manager Quality Assurance	*Darryl Docter*
DATE	02-13-14	

ClearGlide™ Wire Pulling Lubricant

IDEAL Industries, Inc.

1. What is the product name for the item listed in the SDS?

2. What is the emergency telephone number listed by the manufacturer to obtain additional information on the product?

Chapter Activity Reading Safety Data Sheets (SDSs)

3. What is the form of the product?

4. What special firefighting procedures are listed?

5. What type of eye protection is required?

6. In the event of eye contact, what emergency first aid procedures should be followed?

Chapter 2

Introduction to the VDV Industry

Voice-data-video (VDV) installations involve the equipment and cabling required to transmit data over distance. A VDV technician must have a knowledge and understanding of the equipment to be installed as well as the tools and test instruments used while performing the installations. While similar to electrical installations, VDV installations are typically low voltage and have their own sets of standards and system characteristics. Also, VDV technology is constantly evolving in order to handle greater quantities of data within shorter time periods.

OBJECTIVES

- List the different VDV standards organizations.
- Describe the differences between the functions of VDV installers and technicians.
- Identify the different systems that require VDV connectivity.
- Identify the changing VDV industry trends.
- Describe VDV cable types.
- List the types of wireless networks.
- Explain the differences between planning, installation, and testing of VDV cabling.

Digital Resources

ATPeResources.com/QuickLinks
Access Code 838502

HISTORY OF TELECOMMUNICATIONS

Telecommunication is the communication of voice, data, images, and video over a distance via electromagnetic or optical sources on one or more hardwired or wireless networks. *Telecommunications* is the field of study involving telecommunication. The Greek root of "tele-" is the combining form of 'far'. It has, over the last couple of decades, come to mean electronic or optical communications. This form of communication was employed for telegraphs and plain old telephone service (POTS). So when it became common for communication to take the form of photos, printed data, voice recordings, and live versions of all of these, the term became a more generic "telecommunication."

The company which originated long-range electronic communications was the American Telephone and Telegraph Company (AT&T). The company traces its roots back to 1885 and was founded by the inventor of the telephone, Alexander Graham Bell. AT&T was a monopoly. They owned the switches, the transmission wires, and the telephones. They controlled long distance and local services, and they employed all switchboard operators, engineers, and technicians. They maintained this monopoly for over 100 years. Finally, in a much publicized case in 1984, the giant company was broken up into the smaller Regional Bell Operating Companies (RBOCs).

After that occurred, consumers could have a local exchange carrier (LEC) for local telephone service and choose from among many different long distance carriers. Also, consumers could purchase their own telephones and were responsible for the phone wires inside homes and businesses. Currently, competitive local exchange carriers (CLECs) compete with LECs to provide VDV services. Examples of CLECs include Google, Verizon, and Comcast®.

Phone wiring was a simple two-wire circuit. AT&T systems had been using a four-wire copper cable for years, which provided users the opportunity to have two phone lines at one location with one cable. In commercial installations, a standard, six-button telephone was the most common type of telephone used. The six-button telephone was a specially designed telephone with a higher capacity cable available only for commercial use. Up to six phone circuits, or lines, were accessible from a single six-button telephone. **See Figure 2-1.** With six-button telephones, a button would light up when the corresponding line was in use. A user could see which lines were available and press an unlit button for a free line to make a call. At least twelve wires in a cable were required to accommodate six-button telephone circuits.

Although computers had been available since the 1970s, there was no uniformity of styles or operating systems. Computers differed between manufacturers. In 1981, IBM introduced the personal computer (PC). The PC made some work processes easier, such as typing, math calculations, and presentations. However, an even greater value was the newly discovered ability to share information without the need for large amounts of physical storage space or the need to transport hardcopy files to other locations.

Networking computers was the result. Through growth and use, digital networks evolved to include office-shared printers, transference of files, and connections with other networks. This led to inter-networking, or the intranet and Internet. The difference between an intranet and the Internet is that an intranet is a private network. An intranet can only be accessed by those who have the authority and the required passwords. It is typically used only for communication between internal users. Email was just the beginning of the Internet. With current technology, all file transmissions are digital. Voice, data, and video files are transmitted via electronic means. **See Figure 2-2.**

Commercial Telephony

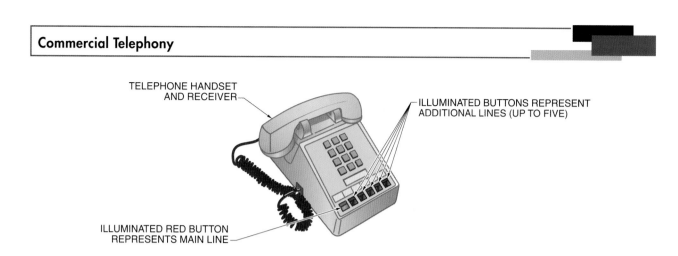

Figure 2-1. In commercial installations, a standard, six-button telephone was the most common type of telephone used. Up to six phone circuits, or lines, were accessible from a single telephone.

Evolution of Personal Computer Hardware and Systems

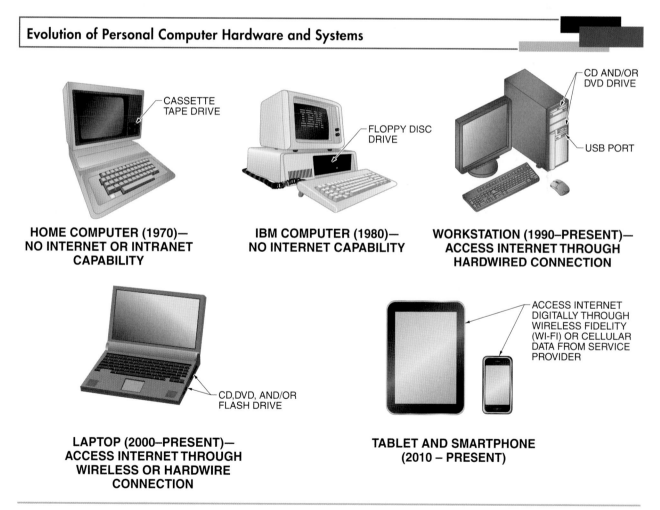

Figure 2-2. The personal computer has evolved from the 1970s home computer to modern devices, such as laptop computers, tablets, and smartphones.

STANDARDS ORGANIZATIONS

As technology evolved, it was determined that there was a lack of understanding of the requirements at the physical layer for the different types of operating systems used for voice and data. For example, the best work area design for both voice and data transmission was not known. Different systems need to have the same operating protocol as another and thereby be able to communicate. In the past, the only company that had been required to design and install these types of systems was AT&T, and they did not do it anymore. However, the technicians who had been designing cabling systems for AT&T were becoming independent consultants known as building industry consultants (BICs). For example, an end user could hire a consultant to design and install a cabling system for an office.

As the need for BICs increased, an organization known as Building Industry Consulting Service International (BICSI) was created in 1977 to bring original equipment manufacturers (OEMs), designers, installers, and distributors together to help standardize installation and design practices.

Building Industry Consulting Service International (BICSI) is a professional telecommunications standards and certification organization. With its headquarters in Tampa, Florida, BICSI has grown to have a membership in over 100 countries. BICSI serves a role in defining standards for the telecommunications industry. BICSI also provides certification for telecommunications design specialists.

Among BICSI's many functions, the organization awards RCDD titles. A *Registered Communications Distribution Designer (RCDD)* is an individual who has acquired education, training, and expertise in the design, implementation, and integration of telecommunications and data transport systems and infrastructure. However, BICSI is not a licensing or governmental body. Legally, a Professional Engineer (PE) is the designation for someone who holds a design license. It is required in most states that electrical drawings must bear the seal of a PE. The need for RCDDs was recognized, and the number of design cabling contractors achieving the RCDD title grew quickly. Having both a PE and an RCDD covers all facets of a project. Today, there are many RCDDs in the United States and around the world, as well as several other credentials available through BICSI.

Tech Tip

In 1996, BICSI recognized the milestone of awarding the 3000th RCDD at the BICSI convention in Kansas City, Kansas.

BICSI also became a standard of training for VDV installers. Being a BICSI-certified installer began to gain popularity in the 1990s. There was, however, another organization which had already been doing electrical installation work for 100 years and had the training schools already in place to handle this new and closely related field. The International Brotherhood of Electrical Workers (IBEW) began to train for and successfully accomplish this work.

STANDARDS

There are several standards that apply to VDV installations. Some standards that apply as well as the organizations that offer them include the following:

- BICSI—Building Industry Consulting Services International provides some training and certifications, but it also publishes guides for proper installation techniques. The *Information Technology Systems Installation Methods Manual (ITSIMM)* is a reference and techniques guide published by BICSI.

- TIA—The Telecommunications Industry Association publishes standards that provide the most widely accepted guidance for proper installation and operation of cabling systems. *Note*: The EIA (Electronics Industry Association) merged with the TIA in 2011.

- ISO—The International Organization for Standardization writes and publishes standards, some of which have a common purpose and subject as the TIA standards, although the TIA standards are more commonly used and cited in the United States.

- NFPA—The *National Fire Protection Association (NFPA)* is a national organization that provides guidance on safety and in assessing the hazards of the products of combustion. The NFPA sponsors and controls the NEC®. The *National Electrical Code® (NEC®)* is a nationally accepted guide regarding the safe installation of electrical conductors and equipment. The distinction of this publication is that it is adopted as law by most states or municipalities. That makes it more than a recommendation; it is illegal to install an electric or communications system in a manner that does not follow these standards. Municipalities enforce these standards by requiring permits and inspections on most projects. **See Figure 2-3.**

INTERNATIONAL BROTHERHOOD OF ELECTRICAL WORKERS

In addition to BICSI, the International Brotherhood of Electrical Workers (IBEW) also provides standards and guidelines for VDV design and installation practices. The *International Brotherhood of Electrical Workers (IBEW)* is a trade organization that represents about 660,000 workers in the electrical, telecommunications, construction, utilities, broadcasting, railroad, manufacturing, and governmental industries. The IBEW was formed in 1891 at a gathering of workers in the new field of electricity. These workers formed the union in St. Louis, Missouri, and quickly spread throughout North America. A decade later, in 1901, the contractors (business owners) formed the management organization known as the National Electrical Contractors Organization (NECA). Together, the IBEW and NECA worked for the safe, effective, and economical design and installation of electrical systems.

In 1941, 50 years after the IBEW came into being, a cooperative effort between NECA and the IBEW created the National Joint Apprenticeship and Training Program (NJATC). The NJATC provides comprehensive and standards-based training for the electrical construction industry. When the need arose for trained technicians, many places around the country began using journeyman electricians to perform the work. Later, a subset of IBEW workers known as "communications technicians" were trained specifically for the voice and data installation field. The NJATC also provided training for these technicians.

In 2014, NECA and the IBEW changed the name of the NJATC to the Electrical Training ALLIANCE. They cited three reasons for this change. First, the new name is more descriptive of what the organization is and what it does. Second, the industry has changed significantly over the past decade or so. There are many new systems and technologies, so it is not the same job, training, or technology that it was previously. Third, the new name serves NECA and the IBEW because it is more easily understood by those not familiar with the extensive program offered.

Tech Tip

The VDV communication field is a technical area that uses terminology involving many acronyms and abbreviations to shorten words and phrases. Understanding VDV acronyms and abbreviations is important when working within the VDV field as they are used on prints and with equipment and test instruments.

The NEC® and Other NFPA Standards

Figure 2-3. It is illegal to install an electrical or communications system in a manner that does not follow the NEC® and NFPA standards.

VDV INSTALLER FUNCTIONS

Electricians properly install conductors, pathways, and electrical equipment. An electrician is trained in termination methods (the connection of wires to connectors), the mandatory requirements of the NEC®, best practice methods for cost-effective installation, and familiarity and safety on job sites. A VDV installer performs similar functions as well. The difference between an electrician and a VDV installer is primarily in the purpose of the installation. An electrician works with systems that transport and deliver electrical power, while a VDV installer works with systems that transport and deliver information electronically.

Because the purpose involves delivering information, not significant power, the actual conductors (wires) used in VDV systems are considerably smaller than those used to transport electric current. The speed of the signal, rather than the voltage level, is the main consideration when specifying VDV conductors. Technology has advanced enough that conductors are not even required to be metallic. VDV systems can now be installed with fiber-optic technology. *A fiber-optic cable is a strand composed of pure glass that is used to carry digital signals, in the form of modulated light, over long distances.* Fiber-optic systems use glass strands encased in cable jackets to conduct light rather than electricity. However, many of the same tools, procedures, termination methods, and occupational safety guidelines apply to both types of technology.

VDV TECHNICIAN FUNCTIONS

VDV technicians install, test, and commission cabling systems and electronics for VDV telecommunications systems. A technician's function can include all the necessary planning for a completed installation including being asked to assemble a bill of materials. Technicians must determine the quantities and types of cable and hardware components needed for specific installations, such as connectors, faceplates, patch panels, etc. There are also technical, aesthetic, and code requirements of which a technician must be aware. Technical requirements involve the following:

- cable type for a system
- twisted-pair equipment requirements
- fiber-optic equipment requirements
- data carrying capacity
- shielded or unshielded cable requirements
- cable termination methods
- connector types

Aesthetic requirements involve the color and material for faceplates; color, type, and size of labels; and types of equipment racks, cable trays, and other visible pathway considerations. Code requirements include adherence to the NEC®; compliance with industry standards, such as those provided by the TIA; and OEM warranty requirements.

There are many other functions that a VDV technician may be required to perform while on the job. In addition to installation of the physical cable plant (all underground and interior cables), a VDV technician often installs electronic components, devices, and related hardware systems. **See Figure 2-4.** At times, a VDV technician may also be required to install software and configure an operating system. For example, video systems require a server, storage space, and a high level of expertise for setup and maintenance. In order to be used, each camera must be identified and operational parameters must be set for how and when to record video.

Commonly Installed VDV Equipment

HARDWARE PLATFORMS

DATA CONTROL CENTERS

CABLING AND RACKS

CAMERAS

TELEPHONE SWITCHING SYSTEMS

COMPUTER SERVERS AND RACKS

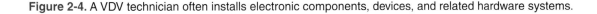

Figure 2-4. A VDV technician often installs electronic components, devices, and related hardware systems.

STRUCTURED CABLING

Structured cabling is a systematic installation of cabling needed for low-voltage systems. Structured cabling is sometimes referred to as "structured wiring" and is installed per selected industry standards. The term "structured cabling" came about because many low-voltage systems use similar cabling. Structured cabling is composed of several subsystems and is interchangeable between different OEMs and upgrades. **See Figure 2-5.**

A *work area* is a location where an individual uses a PC, a telephone, and other user equipment devices connected to structured (horizontal) cabling. *Horizontal cabling* is the cabling that runs from a work area outlet (WAO) to a telecommunications closet. Horizontal cabling is usually a four-pair copper cable, but it can be a fiber-optic cable.

A work area has an WAO to connect fixed VDV devices to a network. A WAO is sometimes called a "telecommunications outlet" or a "jack," and is located on a wall or floor, similar to an electrical outlet.

Backbone cabling is telecommunications cabling and hardware that runs from a telecommunications closet, or intermediate distribution frame (IDF), to a main distribution frame (MDF) or to another telecommunications room. There can be several telecommunications closets in a building but usually only one MDF. Backbone cabling is sometimes referred to as "riser cabling."

Tech Tip
Structured cabling within an institution, such as a school, hospital, or office campus, is premise cabling.

Structured Cabling

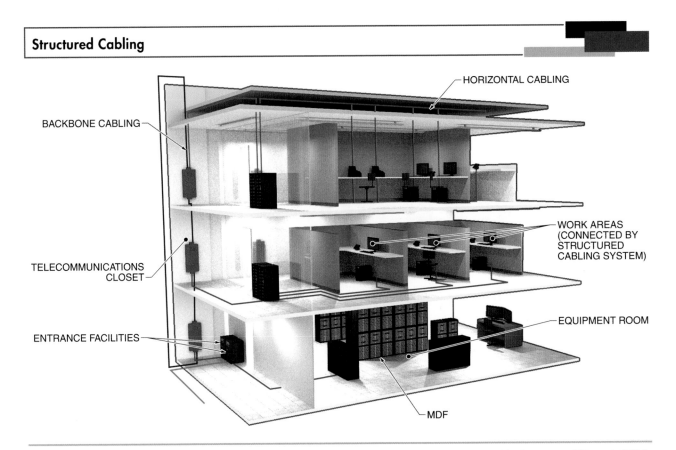

Figure 2-5. Structured cabling is composed of several subsystems and is interchangeable between different OEMs and upgrades.

A *service entrance facility* is the location where telecommunications cabling enters a building. Sometimes called a "demarcation point," it may be located in the MDF but is often located at the nearest corner of a building to the outside incoming service. For this reason, a technician will often be asked to provide a demarcation extension which extends it to the MDF.

Many institutions use campus cabling, such as hospitals, schools, and commercial businesses with multiple buildings in one location. *Campus cabling* is telecommunications premise cabling that interconnects buildings via underground or overhead cabling means. Campus cabling operates over a fairly short distance and does not obstruct public rights-of-way, such as roadways, public sidewalks, etc.

OTHER SYSTEMS AND CABLING

There are many other, closely related systems that commonly fall within the responsibility of a VDV technician to install. These other systems may include security systems, fire alarm systems, sound systems, audio-visual systems, and specialty systems.

Security Systems

Security systems fall into two categories—surveillance systems and access control systems. Security systems are typically used in commercial businesses and institutions. However, many modern residential installations also include security systems. **See Figure 2-6.**

The functions of a surveillance system and an access control system are unique. Due to customers' need for ease of use, many software manufacturers now offer integrated systems that can share data and even control both systems from a common platform. For example, an integrated system can detect when a door is opened and then use a camera to record an image of the one who opened it.

Access Control Systems. An *access control system* is a security system that controls door locks. These systems are independently cabled and separate from other structured cabling systems in a facility. Although most now use a common computer for a control station, and therefore use an Ethernet (LAN) connection via a data network, the individual devices have dedicated and specific cables. Commonly installed access control systems include door locks that are controlled by card readers, biometrics (such as a user's fingerprint, iris-scan, or handprint), push-to-exit panels, and communication pads.

Surveillance Systems. A *surveillance system* is a system that controls security cameras. Most legacy (pre-Internet) surveillance camera systems were connected with standard coaxial cables. Today, most surveillance cameras are network-ready cameras or Internet protocol (IP) cameras. Usually referred to as just "IP cameras," they use the same four-pair cable as a data network. This is because the IP signaling used is the same as that used with a computer, which allows new cameras to use the same types of switches, cabling, patching, etc. as a data network. However, because the volume of data involved with a video signal is so much greater than the amount of data involved with other systems, the network used for video systems is usually separate from the data network.

Access control systems allow users to enter a locked building through electronic control of door locks.

Security Systems

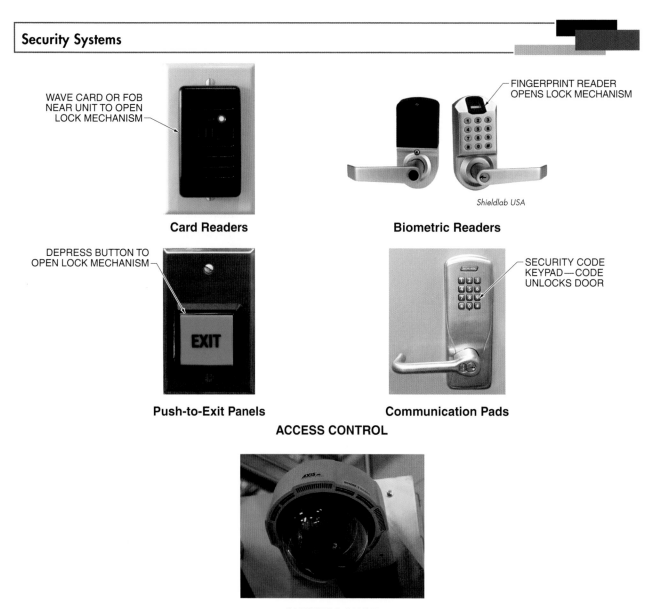

WAVE CARD OR FOB
NEAR UNIT TO OPEN
LOCK MECHANISM

Card Readers

FINGERPRINT READER
OPENS LOCK MECHANISM

Shieldlab USA

Biometric Readers

DEPRESS BUTTON TO
OPEN LOCK MECHANISM

EXIT

Push-to-Exit Panels

SECURITY CODE
KEYPAD—CODE
UNLOCKS DOOR

Communication Pads

ACCESS CONTROL

SURVEILLANCE

Figure 2-6. Commonly installed security systems include access control and surveillance systems.

Fire Alarm Systems

A *fire alarm system* is a low-voltage system used to notify building occupants of a possible fire upon the detection of smoke, heat, or both. Fire alarm systems are code driven, which means that the installer must follow stricter guidelines when installing, programming, and testing a system. All installed fire alarm systems are approved and inspected by an authority having jurisdiction (AHJ). An *authority having jurisdiction (AHJ)* is a person who has the delegated authority to determine, mandate, and enforce code requirements established by jurisdictional governing bodies. Fire alarms have dedicated cables, devices, control panels, and communications systems for the announcement and reporting of a detected smoke or fire event.

Sound Systems

A *sound system* is a network of hardwired or wireless speakers installed throughout a building to transmit sound only, such as music, alarm notifications, or paging, throughout a specified location. Sound systems vary in type and use. For example, many commercial and industrial offices have some basic type of public address system. The system can be as simple as a number of speakers mounted in a ceiling at regular intervals. It can also be a more complicated system that provides white noise for privacy, music, programming, or other purposes. In theaters, sound systems are complex and more powerful. A VDV technician may have to install the cabling, speakers, electronics, microphones, sensors, and other components for these systems.

Audio-Visual Systems

An *audio-visual (AV) system* is a network of hardwired or wireless video monitors that transmits sound and video, usually for presentations to large groups of people located in a common area. AV systems are often used for educational or communications purposes. From a video conferencing system installed in a corporate boardroom to the smart boards installed in many classrooms, AV systems are growing in use and complexity. These systems involve the integration of displays, computers, cameras, projectors, and almost any type of visible or audible data.

Specialty Systems

There are types of specialty systems that use specific, low-voltage cables and electronics. **See Figure 2-7.** A few common specialty systems include the following:

- hospitals—infant abduction systems
- hospitals—patient data and medical record systems
- casinos—slot machine data systems
- retail—point-of-sales (POS) systems
- warehouse and retail—inventory control systems
- in-building wireless phone systems
- distributed antenna systems
- emergency public address systems
- lED lighting systems

Other Systems

FIRE ALARM CONTROL PANELS

FIRE ALARM

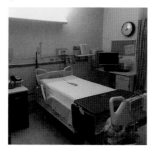

SOUND

AUDIO-VISUAL (AV)

Hospitals

Retail

SPECIALTY

Figure 2-7. Other VDV systems that are commonly installed and serviced by VDV technicians include fire alarm systems, sound systems, AV systems, and specialty systems.

SYSTEM INTEGRATION

As computing power has increased, so has the ability to integrate systems. For example, it is common for a card reader on an access control system to initiate a camera recording, even if they are components of two different systems. This is accomplished by interconnecting multiple systems through cabling or shared controllers as well as through the configuration of the software. Any system with a detection ability, such as a door contact, a motion detector, a card reader, or a smoke detector, can be used to trigger another system. That can include sending an email, calling an authority, recording sound or video, or simply logging the event in an electronic file.

CHANGING INDUSTRY TRENDS

The industry is always changing due to the need to increase transmission speeds. Greater amounts of information are searched, analyzed, and presented to more users than ever before. Users have come to expect instant information, directions, and opinions about nearly every topic imaginable. The speed of transmission and access is making this possible. Huge data centers have been built and more are being built to accommodate the sheer volume of information transmitted. Cable types have been designed to accommodate changing industry trends and wireless networks.

Increasing Speed

In the last 20 years, common network speeds have gone from 1 megabit per second (Mbps) to 10 gigabits per second (Gbps). There is also ongoing research to increase speeds to create even faster networks for the purpose of moving greater amounts of data. For example, systems that use videos, maps, movies, and medical imaging all involve greater amounts of data, and the user expects nearly instant delivery of that data.

Video Content Analytics

A VDV system can use video content analytics (VCA) to analyze video input and perform a function, such as sound an alarm, when a change is detected. This is accomplished with algorithms that the system uses to identify and evaluate changes as they occur. An algorithm is a finite set of well-defined rules or processes that can be used to solve a problem. VDV systems are computer-controlled and can make decisions based upon what is detected. For example, software exists to ignore the presence of a person in a camera's field of view. However, the camera may begin recording

and send a notification if that person moves or leaves an object, such as a briefcase, and walks away from it. This type of processing requires fast action and the equipment to support it.

Cable Types

There are two standard types of VDV cabling—copper cable and fiber-optic cable. Both types are further classified into subgroups. VDV cable is often labeled using the acronym of the cable, such as unshielded twisted pair (UTP), screened twisted pair (ScTP), or shielded twisted pair (STP). ScTP and STP have different types of shielding from electromagnetic interference but are otherwise similar. With copper cables, the most common type is UTP, which is available in several performance categories. In addition, these cable types are changing further to accommodate greater amounts of data at faster speeds.

With coaxial cable, an outer braided conductor is wrapped around an insulated inner conductor. The outer conductor is used to shield the inner conductor from outside electromagnetic interference (noise).

Current state-of-the-art cabling most commonly installed in new or updated installations is Category (Cat) 6A, which is an enhanced level 6 cable that has the highest performance level (data carrying capacity) for UTP and ScTP cables. Other types of copper cabling include coaxial cable, used mostly for video transmission, and STP, which provides increased bandwidth and stronger signal security than UTP. With fiber-optic cable, there are similar subgroups based on performance level. The higher the quality of the cable, the faster the speed of data transmission.

Conductors used for VDV systems are available as individual wires or in groups, such as with cables and cords. **See Figure 2-8.** A *wire* is an individual conductor. A *cable* is a group of two or more conductors within a common protective cover and is used to connect individual components. A patch cord is an insulated copper or fiber-optic cable used to connect two devices or components by means of a plug at both ends.

VDV Cable Types

4 UTP	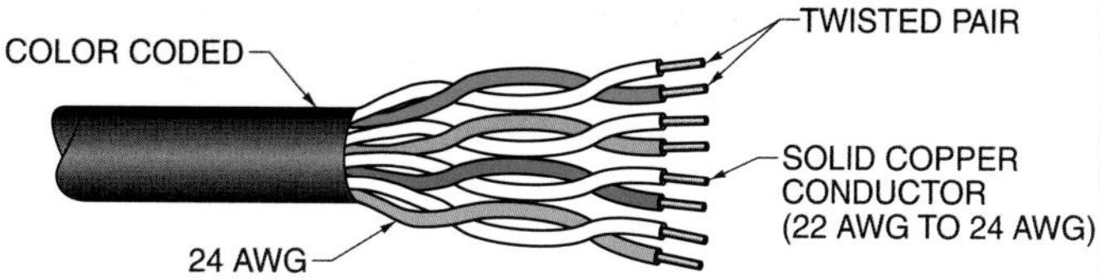 COLOR CODED — / TWISTED PAIR / SOLID COPPER CONDUCTOR (22 AWG TO 24 AWG) / 24 AWG	TWISTED-PAIR CABLE USED FOR VOICE AND DATA TRANSMISSION
25 UTP	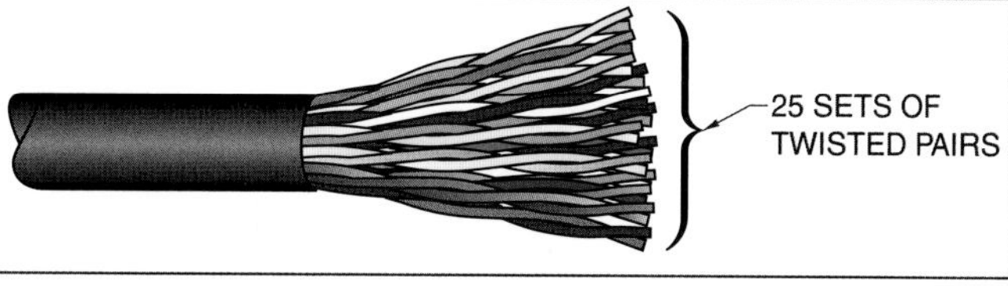 25 SETS OF TWISTED PAIRS	CABLE USED FOR DATA TRANSMISSION
COAXIAL CABLES	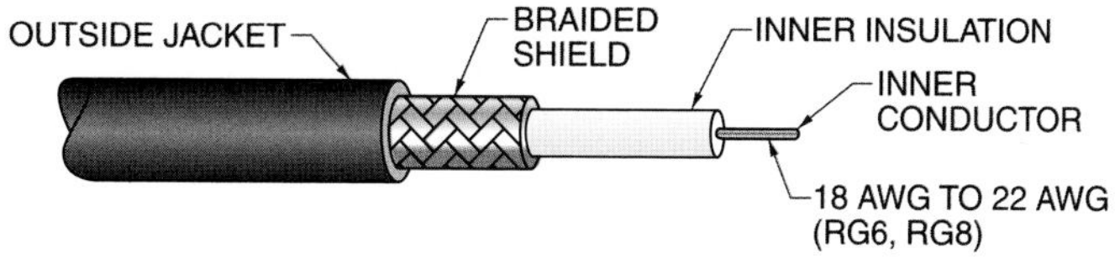 OUTSIDE JACKET — / BRAIDED SHIELD / INNER INSULATION / INNER CONDUCTOR / 18 AWG TO 22 AWG (RG6, RG8)	CONNECTS CABLE TELEVISION AND VCR TO TELEVISION; ALSO USED FOR DATA
CAT 3/FTP/ 100PR/24 AWG	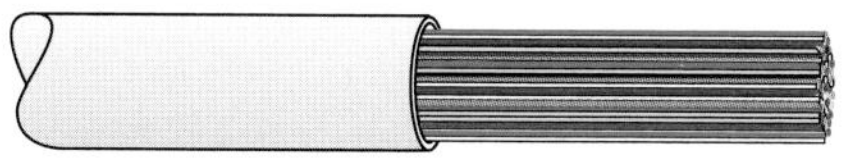	CAT 3 ISO 11801/TIA 568A COMPLIANT. USED AS BACKBONE CONNECTION FOR DATA SYSTEMS
ALUMINUM ARMOR OSP CABLES	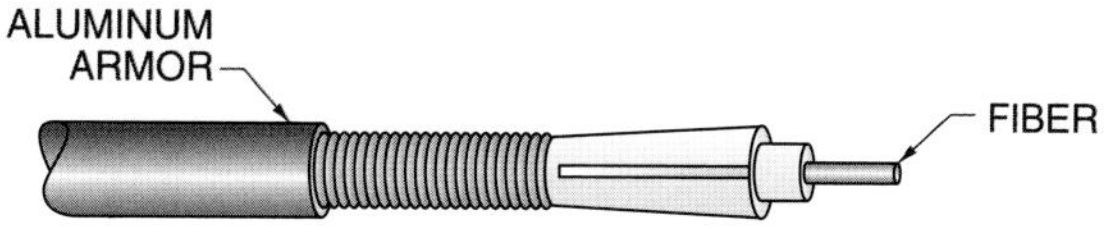 ALUMINUM ARMOR — / FIBER	CAT 3 ISO 11801/TIA 568A COMPLIANT. USED AS BACKBONE CONNECTION FOR DATA SYSTEMS
UTP CABLES	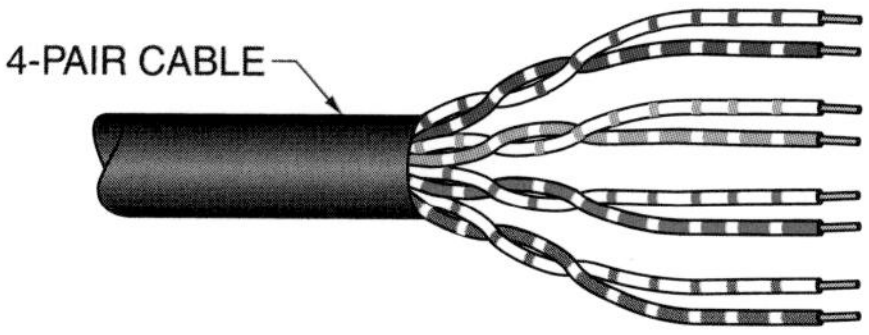 4-PAIR CABLE —	TWISTED-PAIR CABLE USED FOR VOICE, DATA, AND VIDEO TRANSMISSION
STP CABLES	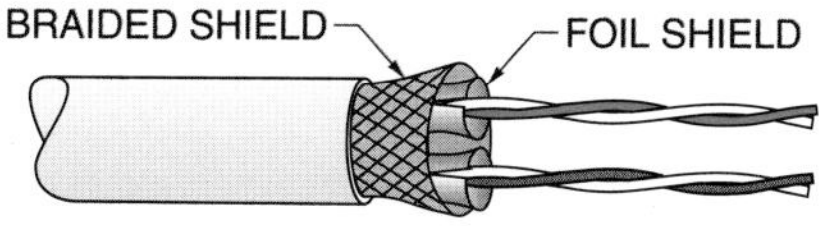 BRAIDED SHIELD — / FOIL SHIELD	USED FOR PROTECTION FROM ELECTROMAGNETIC INTERFERANCE
ScTP CABLES	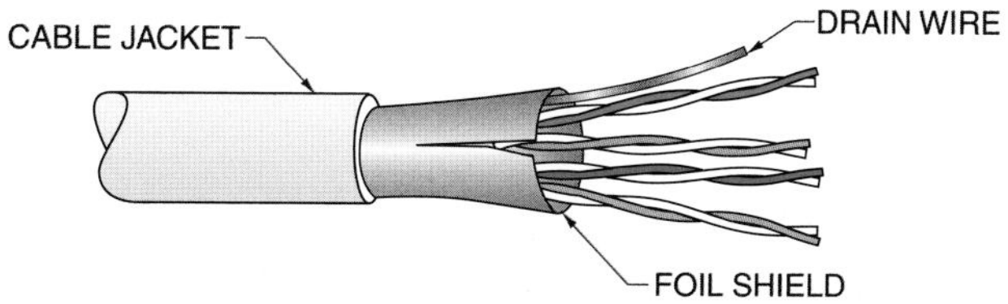 CABLE JACKET — / DRAIN WIRE / FOIL SHIELD	USED FOR PROTECTION FROM ELECTROMAGNETIC INTERFERENCE; HAS LESS RESISTANCE THAN STP
FIBER-OPTIC CABLES	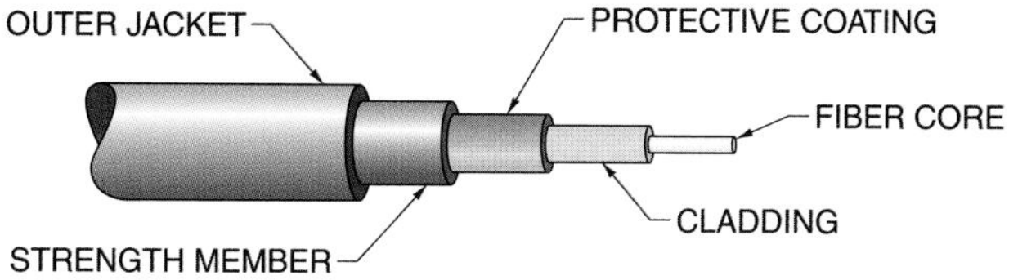 OUTER JACKET — / PROTECTIVE COATING / FIBER CORE / CLADDING / STRENGTH MEMBER —	FIBER-OPTIC CABLE USED OUTSIDE FOR DIRECT BURIAL AND LONG RUNS

Figure 2-8. VDV cable types include cables and cords.

Most individual conductors are enclosed in an insulated cover to protect the conductor from damage and wear, to protect the user from electric shock, and to meet code requirements. Some individual conductors, such as a ground wire, may be bare.

Wireless Networks

In the 1970s, the sound industry used the term "hi-fi" which was an abbreviation for "high fidelity sound reproduction." High fidelity is the reproduction of sound with a high degree of faithfulness to the original. It was memorable and easily understood by users that "hi-fi" meant "good sound."

A *wireless access point (WAP)* is a device that connects wireless telecommunication devices to form a wireless network. Devices connected are typically desktop computers, laptop computers, and controllers. A *controller* is a device used to control another piece of VDV equipment. With the advent of WAPs, the industry borrowed a term from the past and called it "Wi-Fi" loosely meaning "wireless fidelity."

In recent years, Wi-Fi has been installed in many commercial and residential buildings. Commercial offices, coffee shops, restaurants, airport terminals, passenger train stations, and even most homes have some form of wireless connection. This has allowed innovations in many areas of technology. For example, new HVAC systems use a Wi-Fi connection to inform the user when to change the air filter. Game systems, cellular telephones, and security systems all use Wi-Fi. The WAPs are antennas for transmitting Ethernet activity. They are easily installed with Cat 5e or better cables. Care in distribution of antennas as well as frequency interference requires skilled VDV technicians to have an understanding of how Wi-Fi operates.

PLANNING, INSTALLATION, AND TESTING OF VDV CABLING

When planning a new cabling system, the VDV technician, with the help of a project manager or a CAD designer, will review the specifications for all VDV systems to be installed. Cable paths are chosen to include the physical layout, capacity, installation environment, and type of support system that is best suited to the full requirements of the specifications. For example, a system of J-hooks can work effectively if there is a low-to-medium volume of cables and little variance in cable types. However, a divided cable tray is more effective for a large cable volume with many different cable types.

Installation practices vary according to the types of cable involved, quantity of locations, local electrical and building codes, and operating environment. Similarly, testing requires specific equipment to be matched to the cable type, level of performance, and type of end point termination used in a cabling system. Making the right choices for these variables can be important for meeting code requirements and for making system maintenance easier.

Summary

The modern VDV industry relies on standards organizations to provide guidelines for proper installation and safety practices. VDV installers and technicians must be familiar with these standards as well as the equipment, cabling, and different systems involved with VDV installations. VDV systems can include security systems, fire alarm systems, sound systems, AV systems, and specialty systems. The VDV technology is constantly evolving to meet the demand for handling greater amounts of data within shorter time periods. VDV technicians and installers must be familiar with the different types of cabling used for such systems as well as the complexities of wireless networks.

Chapter Review

1. What is backbone cabling?

2. What is another term for backbone cabling?

3. What is an RCDD?

4. Explain the difference between an electrician and a VDV technician.

5. Which was the first company to originate long-range electronic communications?

6. List four organizations that provide standards for the VDV industry.

7. List six technical requirements associated with VDV equipment.

8. What are the two categories of security systems?

9. List four commonly installed components of access control systems.

10. What is POTS?

11. List five types of VDV specialty systems.

12. What are the two main types of VDV cable?

Chapter Review

13. What is the most commonly used type of copper cabling?

14. What are two devices that transmit a Wi-Fi signal to user equipment?

15. Why was BICSI created?

Chapter Activity VDV Cable Identification

Identify each cable shown.

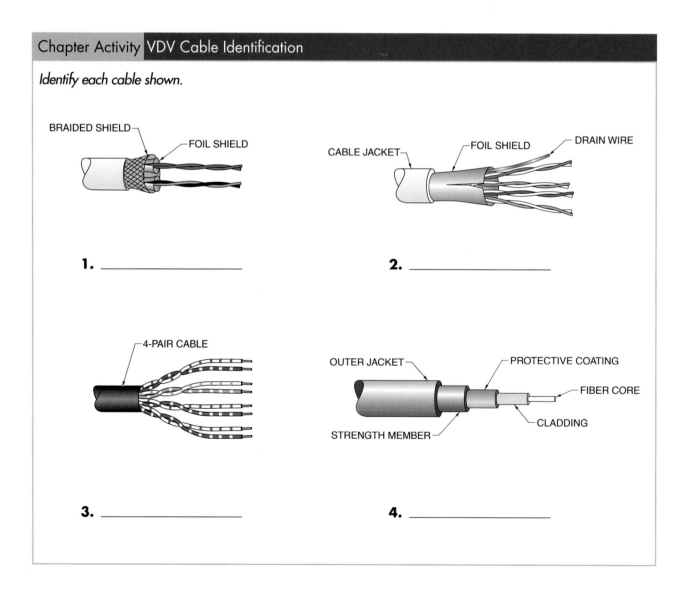

BRAIDED SHIELD
FOIL SHIELD

1. _____

CABLE JACKET
FOIL SHIELD
DRAIN WIRE

2. _____

4-PAIR CABLE

3. _____

OUTER JACKET
PROTECTIVE COATING
FIBER CORE
CLADDING
STRENGTH MEMBER

4. _____

Chapter Activity VDV Cable Identification

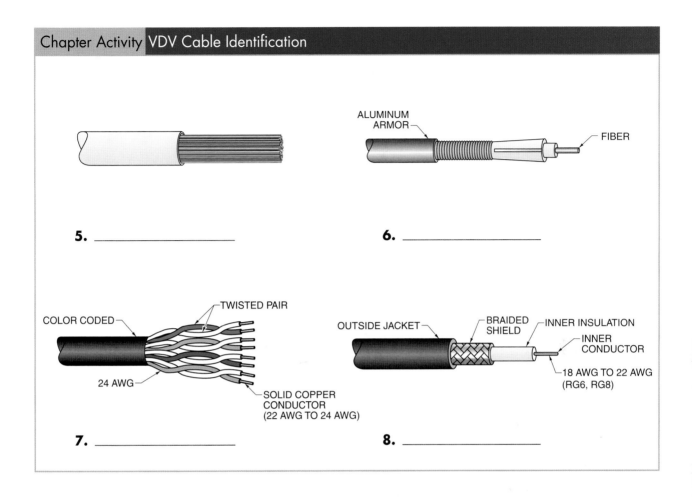

ALUMINUM ARMOR

FIBER

5. _____

6. _____

COLOR CODED

TWISTED PAIR

24 AWG

SOLID COPPER CONDUCTOR (22 AWG TO 24 AWG)

OUTSIDE JACKET

BRAIDED SHIELD

INNER INSULATION

INNER CONDUCTOR

18 AWG TO 22 AWG (RG6, RG8)

7. _____

8. _____

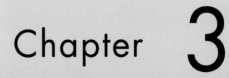

Chapter 3

VDV Tools

Every trade, including the voice-data-visual trade, requires the use of specific tools. Because technology is changing rapidly, there are many different types and manufacturers of systems, cabling, and tools available for VDV installation, maintenance, and troubleshooting. Many tools and test devices used for VDV applications are similar to more common tools and test equipment, but they come with modifications and are designed for use specifically with VDV systems.

OBJECTIVES

- Explain the proper use of tools and test equipment for the installation and troubleshooting of VDV equipment.
- List the steps for proper safety procedures when working with VDV tools and test equipment.
- Describe the proper application of fiber-optic test equipment such as fusion splicers, optical attenuation test sets, and optical time domain reflectometers.
- Explain how electronics are used with modern VDV test equipment.

Digital Resources

ATPeResources.com/QuickLinks
Access Code 838502

HAND TOOLS

Most of the hand tools used by VDV technicians are common hand tools found in many toolboxes and can be used for many different tasks. **See Figure 3-1.** Hand tools used by VDV technicians include the following:

- toolbox or bag
- cable cutters
- long-nose pliers
- tongue-and-groove pliers
- wire stripper
- hammer
- 6″ adjustable wrench
- utility knife
- hacksaw
- tape measure
- fish tape
- level
- electrician's scissors
- punchdown tool
- glow rod
- VDV cable stripper
- set of screwdrivers
- drywall saw
- flashlight
- tone generator

Most of the listed tools are commonly used in any trade. However, pairs of electrician's scissors, punchdown tools, fish tapes, and glow rods are hand tools that are typically used in the electrical and VDV trades.

Electrician's Scissors

Electrician's scissors are specialized scissors that, in addition to having a notched edge for stripping cable insulation, are sharp enough to cut aramid fibers. *Aramid* is a high-strength, flexible, heat-resistant synthetic fiber. Aramid is better known by the trade name Kevlar®. Kevlar is typically used as fabric for bulletproof body armor, in tires, or as an asbestos substitute.

Because of its good tensile strength, flexibility, and light weight, aramid fiber is also woven into yarn and used with fiber-optic cabling to strengthen it. Aramid yarn prevents fiber-optic cables from being overbent and kinked.

When terminating or installing cable, the aramid yarn must be cut cleanly. Clean cuts to aramid yarn can only be made with electrician's scissors rather than standard scissors.

Punchdown Tools

Another type of hand tool that is typically only used in VDV applications is a punchdown tool. This tool is used to terminate copper cabling onto many types of termination and connection hardware. **See Figure 3-2.**

A *patch panel* is a device used to make connections between incoming and outgoing VDV lines. Patch panels can be fully loaded with RJ-45 jacks on the front and 110 termination blocks on the back. Patch panels can also be a simple frame to hold individual jacks for RJ-45 connectors. A *patch cord* is a flexible 3′ to 12′ length of cable used to connect a network device to a main cable run or a panel. An *RJ-45 connector* is an 8-pin connector used for data transmission over standard telephone wire. RJ-45 jacks also have 110 termination hardware built into them. A *110-type cross-connect* is a compact connecting device that can be arranged for use with either jumper wires or patch cords. For voice applications, older 66 blocks, used to terminate and cross-connect twisted-pair cables, or newer 110 blocks are used for cross-connecting switches to individual phone lines.

Both 66 and 110 blocks are insulation displacement connecting (IDC) blocks, and punchdown tools are used to terminate insulated copper wiring on these blocks. They are commonly used on patch panels, connectors, and wall termination cross-connection blocks. Manufacturers of data transmission systems prefer the more modern 110 blocks because 110 blocks are more compact with lower noise than the less modern 66 blocks. Older analog voice systems sometimes still have 66 blocks.

Common VDV Hand Tools

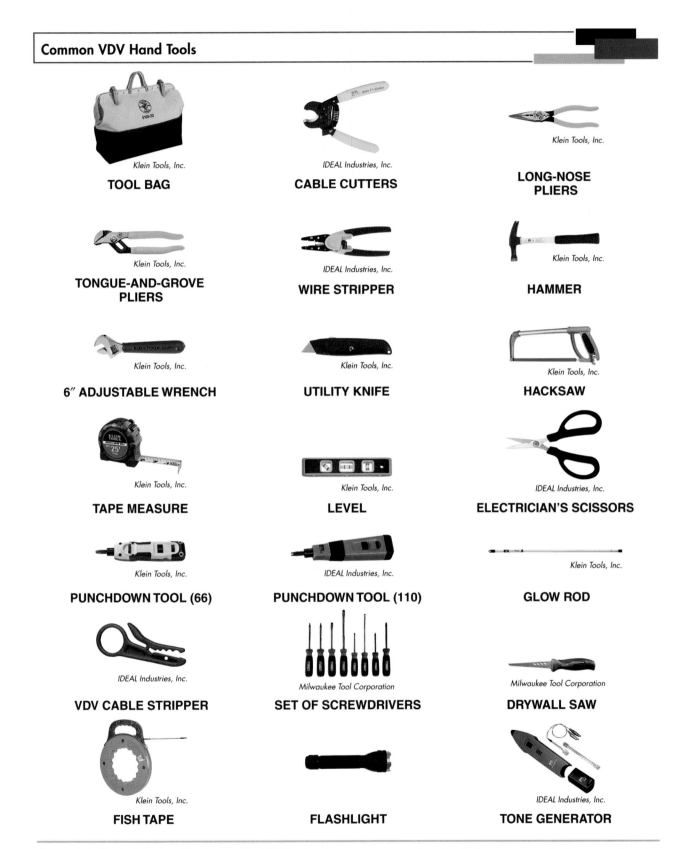

TOOL BAG — Klein Tools, Inc.

CABLE CUTTERS — IDEAL Industries, Inc.

LONG-NOSE PLIERS — Klein Tools, Inc.

TONGUE-AND-GROVE PLIERS — Klein Tools, Inc.

WIRE STRIPPER — IDEAL Industries, Inc.

HAMMER — Klein Tools, Inc.

6" ADJUSTABLE WRENCH — Klein Tools, Inc.

UTILITY KNIFE — Klein Tools, Inc.

HACKSAW — Klein Tools, Inc.

TAPE MEASURE — Klein Tools, Inc.

LEVEL — Klein Tools, Inc.

ELECTRICIAN'S SCISSORS — IDEAL Industries, Inc.

PUNCHDOWN TOOL (66) — Klein Tools, Inc.

PUNCHDOWN TOOL (110) — IDEAL Industries, Inc.

GLOW ROD — Klein Tools, Inc.

VDV CABLE STRIPPER — IDEAL Industries, Inc.

SET OF SCREWDRIVERS — Milwaukee Tool Corporation

DRYWALL SAW — Milwaukee Tool Corporation

FISH TAPE — Klein Tools, Inc.

FLASHLIGHT

TONE GENERATOR — IDEAL Industries, Inc.

Figure 3-1. Most of the hand tools used by VDV technicians are common hand tools found in many toolboxes and can be used for many different tasks.

Punchdown Tools

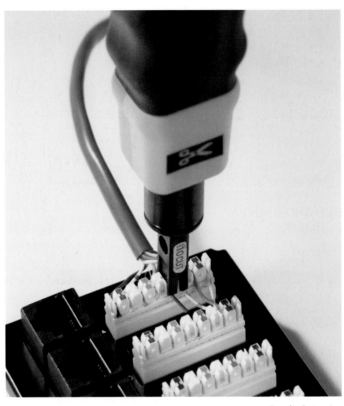

Greenlee Textron, Inc.

Figure 3-2. A punchdown tool is used to terminate copper cabling connectors.

Voice over Internet Protocol (VoIP) is the method and technology used to transmit voice communication over a data network through use of the Internet. Modern VoIP systems are actually voice software packages operating on computer networks. However, there are many analog and digital phone systems still in use that require the use of punchdown tools. The copper cable used for voice and data applications is typically a 24-gauge, multipair cable.

Prior to using punchdown tools for termination procedures, wires were stripped of their insulation and wound around a screw or a metal post. This method was slow, tedious, and resulted in too much variance among terminations. This method was used when installing old (pre-Internet era) legacy systems, but it did not work well when installing modern high-speed data networks.

Note: Currently, almost all copper cable for data systems is terminated via 110 terminations. There are other types for fiber-optic systems and older legacy terminations that are still used, but almost all new installations use a 110 IDC.

Knowledge of legacy termination methods is required for VDV technicians since older installations may have legacy technology. There are applications involving modern systems, such as access control systems or fire alarm systems, which may interface with a VDV network and require the use of one of these termination methods. **See Figure 3-3.**

Legacy Termination Methods

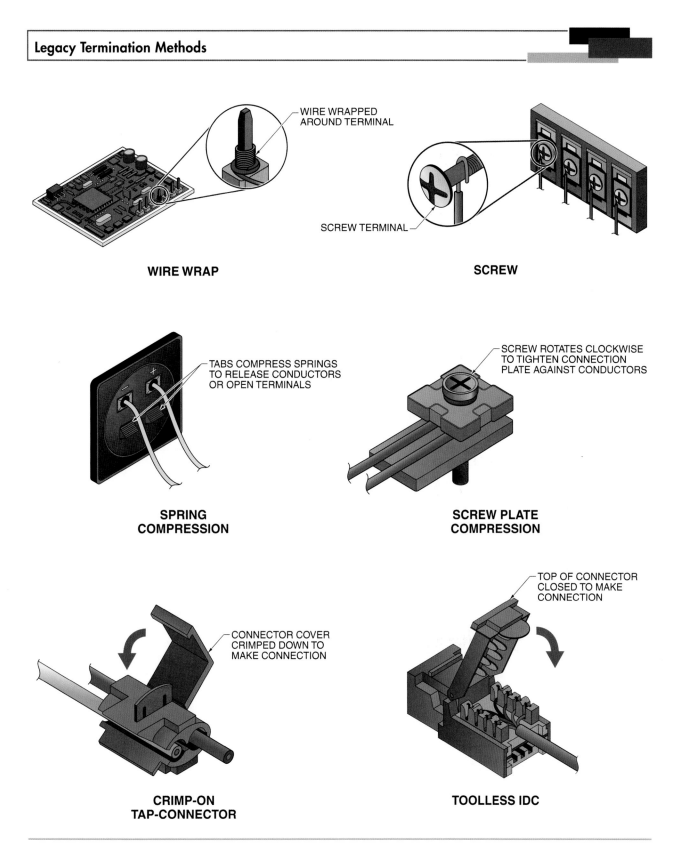

Figure 3-3. Legacy terminations may be used in applications involving modern systems that interface with a VDV network, such as access control systems or fire alarm systems.

A punchdown tool is used to perform a fast termination method referred to as "insulation displacement." Punchdown tools are manufactured by several different OEMs and have interchangeable blades intended for use with specific types of termination blocks.

The insulation displacement process with a punchdown tool increases the speed of making terminations and provides a consistent, repeatable result. The tool works by placing an individual wire in a groove between two guide posts of a 110 termination block and then piercing the bare section of wire. This "punchdown process" seats the wire in place and trims off excess wire. The metal contacts are positioned to cut through the insulation and make contact with the copper wire inside. The conductor is evenly wedged between the contacts to hold it securely in place. **See Figure 3-4.**

Fish Tape and Glow Rods

A *fish tape* is a flexible cable that is used to move another cable through a tight space that cannot be reached. A *glow rod* is a semirigid rod that is used to move a cable through a space that cannot be reached.

A fish tape is stored on a reel and pulled out to the length needed to feed it through a tight space, such as a raceway, crawl space, or behind drywall. Once the end of the fish tape is completely fed through the space, a cable or conductor is attached to a hook or an eyelet at the end of the fish tape and pulled back through the space.

Tech Tip

Several accessories are available to place on a glow rod tip to help push the rod over any obstructions in a passageway.

Insulation Displacement

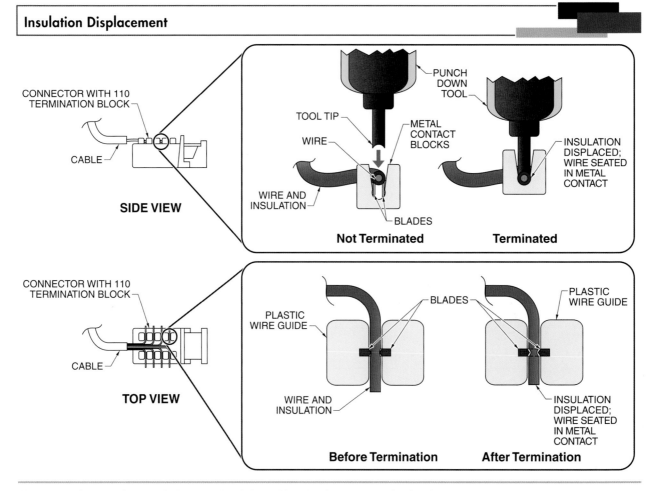

Figure 3-4. A punchdown tool displaces insulation without having to cut and strip wires separately.

A glow rod has a diameter roughly equivalent to that of a pencil and telescopes out to a length ranging from 6′ to 35′. It serves the same purpose as a fish tape, but it is typically used in situations when a space is visible but cannot be reached through ordinary means. The term "glow rod" comes from its luminescent property to glow in the dark. The luminescence from a glow rod provides enough light to see it as it is fed through a space or to locate it in a dark space, such as an attic or above a lay-in ceiling.

TEST EQUIPMENT

All test equipment must be used per the written instructions from the OEM. Every OEM has instructions that are included with their equipment when purchased. However, it is recommended to view the instructions on the OEM's website as well, since the latest product updates and revisions to instructions are typically posted online. **See Figure 3-5.**

Wiremap Testers

A *wiremap test* is a cable test used to determine the quality of the connection between both ends of a cable. Wiremap tests are also used to test for the absence of shorts, grounding, and external voltage. A *wiremap tester,* or verification tester, is a test instrument used to locate open circuits, short circuits, and improper cable terminations such as swapped conductors. A wiremap tester typically displays a "PASS" or "FAIL" indication of the wire's condition within all four pairs of the cable. It tests and displays the wire conditions between the tester ends of the cable and the far ends of the cable on all four pairs. **See Figure 3-6.**

Wiremap Testers

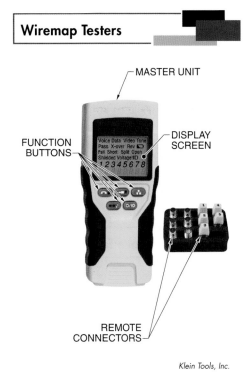

Klein Tools, Inc.

Figure 3-6. A wiremap tester is a test instrument used to locate open circuits, short circuits, and improper cable wiring.

Common VDV Test Equipment OEMs
Company Name
Fluke Networks
Agilent Technologies
Corning, Inc.
Harris Corp.
Ideal Industries, Inc.
JDS Uniphase Corp.
Klein Tools, Inc.
Lumentum

Figure 3-5. There are several original equipment manufacturers (OEMs) of VDV test equipment.

As with most tools, several OEMs offer wiremap testers. A wiremap tool is an inexpensive tool which quickly verifies if the terminations on an RJ-45 connector are in the correct position. Usually, there are two components—a "master" and a "slave." The master is the end that displays the test data, records the test data, or both. The slave is a reference point for the test and has no display or controls.

A common RJ-45 connector has eight connections (or pins) and is used for data transmission. A twisted-pair cable also has eight conductors, configured in pairs (four pairs) in one cable. The output of the tester simply indicates if the eight pins on one end of the cable are properly connected to the corresponding pins on the opposite end of the cable. This includes the identification of short circuits between pairs, crossed pairs, reversed pairs, and split pairs.

TSB-67 Compliance Testers

Transmission Performance Specifications for Field Testing of UTP Cabling Systems (TSB-67) is a document that specifies the electrical characteristics of field test instruments, test methods, and minimum transmission requirements for UTP cabling. Some OEMs refer to a TSB-67 compliance tester as a cable analyzer. Most OEM warranties, end users, or system designers typically require this test to certify any Cat 5 or above cabling installation. **See Figure 3-7.**

As with a wiremap tester, a TSB-67 compliance tester also has a "master-and-slave" configuration. However, it measures more parameters than a wiremap tester. As per *TSB-67,* a TSB-67 compliance tester provides a wiremap test and also measures twisted-pair length, attenuation, near-end crosstalk (NEXT), far-end crosstalk (FEXT), and attenuation to crosstalk ratio far-end (ACR).

TSB-67 Compliance Testers

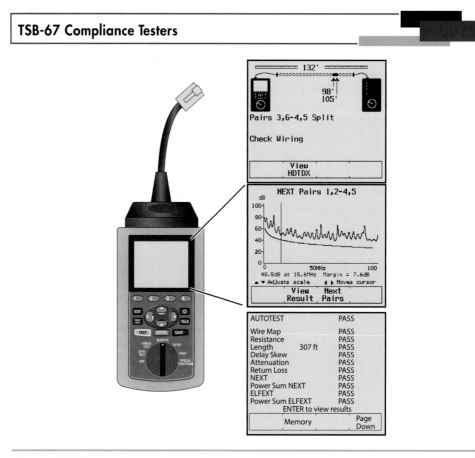

Figure 3-7. TSB-67 compliance testers are similar to wiremap testers but also measure length by electrical means, attenuation, and near-end crosstalk (NEXT).

A *wiremap (cabling map)* is a map that shows how VDV cabling is routed throughout a building. A wiremap helps determine the true length of cable because the twist of pairs within a cable makes the measured copper length longer than the footage marked on the outside cable sheath. Because all pairs are not twisted at the same rate, there is a difference in length from one pair to the next. *Attenuation* is the reduction of power in any signal, light beam, or light wave, either completely or as a percentage of a reference value. Attenuation is also known as the measure of signal lost in the cable length.

Crosstalk is a type of interference caused by signals from one circuit into adjacent circuits. *Near-end crosstalk (NEXT)* is the measure of the amount of signal interference from one pair into the pair next to it in the same cable. NEXT can typically be controlled through the use of shielded or screened cabling systems. *Far-end crosstalk (FEXT)* is the measure of signal interference at the transmit end of a twisted-pair cable from a transmit pair to an adjacent pair, as measured at the far end of the cable. NEXT and FEXT are caused by poor terminations, crushed or touching conductors, or both. *Attenuation to crosstalk ratio (ACR)* is the measure (ratio) of signal loss (attenuation) to near-end crosstalk.

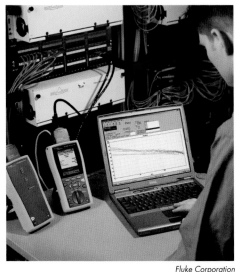

Test instruments such as cable analyzers can be used with computer software to determine the quality of installed structured cabling.

Butt Sets

When working on plain old telephone systems (POTS), a technician often uses a special type of test equipment known as a butt set. A butt set is sometimes referred to as a telephone test set and is a functional analog telephone. **See Figure 3-8.** The term "butt set" comes from the use of this equipment allowing technicians to "butt in" on phone calls. Technicians use butt sets to patch into telephone lines to verify that they are in proper working order as well as to run computerized tests on the lines.

In the past, it was common practice for a technician to patch into a telephone line to place a call. However, cellular telephones have all but eliminated this practice.

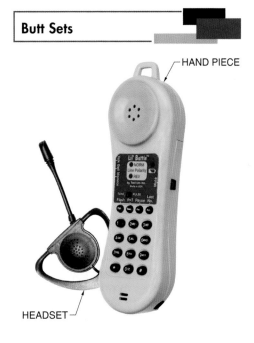

Butt Sets

— HAND PIECE

HEADSET ╯

Figure 3-8. A butt set is sometimes referred to as a telephone test set and is a functional analog telephone used for working on plain old telephone systems (POTS).

Fiber-Optic Termination Kits

A *fiber-optic termination kit* is a tool kit used by a VDV technician that contains all of the necessary tools for terminating fiber-optic pre-polished and field-polished connectors.

Typically, a fiber-optic termination kit is self-contained in a hard- or soft-sided carrying case to keep the tools and components separated from other types of equipment used by the technician. **See Figure 3-9.**

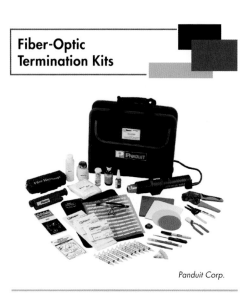

Fiber-Optic Termination Kits

Panduit Corp.

Figure 3-9. A fiber-optic termination kit is a tool kit that contains all of the necessary tools for terminating fiber-optic prepolished and field-polished connectors.

Since cable termination methods have evolved greatly over the last few years, the tools included in a standard fiber-optic termination kit have likewise evolved. Currently, most fiber-optic terminations involve crimp-on connectors. A fiber-optic termination kit usually contains some variation of the following tools:

- fiber cleaver—used to evenly cut a glass-fiber strand
- crimp tool—used to lock a connector onto a fiber strand
- fiber stripping tool—used to strip the outer coating off of a glass fiber
- cable stripping tool—used to strip or cut inner buffer tubes
- fiber cleaning solution—used to clean foreign substances off glass-fiber strands
- wipes—used to remove excess cleaning solution

- electrician's scissors—used to cut through aramid yarn, cable jackets, and buffer tubes
- marker
- guide—used for cutting lengths
- safety glasses—used to prevent glass from entering the eyes

Connector types other than crimp-on types are still occasionally used and usually require different methods of securing an outer connector housing to a glass strand. For example, some types use adhesive which is cured either by heat or UV light. For connectors that require this type of termination method, the connector kit would contain an oven or a UV light source to cure the adhesive as well as polishing pads and a fiber scope.

A *fiber scope* is a device with a magnifying lens on one end and a fiber-optic connection on the other. A fiber scope allows a technician to inspect the end of a polished connector and visually confirm the quality of the termination. Connectors that are not of the crimp-on design require a technician to hand polish a connection. These types are becoming obsolete due to the much longer installation time involved as well as the material waste generated that is required for termination.

Fiber-Optic Fusion Splicers

A *fusion splice* is a permanent splice accomplished by the application of heat with a temperature high enough to melt the ends of two sections of glass fiber. A *fiber-optic fusion splicer* is a tool used to weld the ends of two strands of fiber into a single length using an electrical arc to melt the ends of the strands. During this process, the glass ends are stripped of all coatings, cleaned of impurities (dirt) on the surface, and positioned precisely for the splice. Modern fusion splicers will align and trigger the fusion automatically. Older versions of fusion splicers require a technician to visually align the fibers and then initiate the process manually. **See Figure 3-10.**

Fiber-Optic Fusion Splicers

Figure 3-10. A fiber-optic fusion splicer is a tool used to weld the ends of two strands of fiber into a single length using an electrical arc to melt the ends of the strands.

Visual Light Sources

A *visual light source* is a light used to verify of the continuity of a fiber strand. It is usually red, because that is the longest wavelength visible to the human eye. A red light beam is also the brightest for a given power level, which makes it the most effective light for visually detecting faults. Although a visual light source is not a laser, care should still be taken not to look directly into the light source.

When connected to one end of a clean fiber, the light should be visible at the opposite end of the fiber strand. If so, it provides a visible confirmation that the fiber is continuous and that the connection is in the correct location in the patch panel at both ends.

Optical Attenuation Test Sets

An optical attenuation test set is sometimes referred to as an optical attenuation tester, a power meter, or a link loss test set.

The measure of how well a fiber-optic link works is signal loss. An optical attenuation test set injects light at a known power level into one end of a fiber strand and then measures the amount of light that is delivered at the opposite end. A simple subtraction of the two levels indicates how much signal loss (attenuation) occurs in the fiber tested. Because this is a direct measurement of loss, it is the most reliable certification test used. The output for each fiber is simply a number measured in decibels (dB) of loss. For example, a typical fiber-optic length with a connector pair and splice might have a total loss of about 1 dB.

A *cable loss report* is an analytical tool that can be created by certain types of test instruments and is usually generated by a VDV technician to certify all of the fibers in a specific cable. Most optical attenuation test sets display, but do not record, the loss level. So a technician is often required to record and save the losses measured. **See Figure 3-11.**

Optical Time Domain Reflectometers

An *optical time domain reflectometer (OTDR)* is a test instrument used to measure fiber-optic cable attenuation. Attenuation is caused by several factors including impurities in the glass, transitions from one fiber to another, or even bends in the cable. The amount of light lost has a direct effect on the data rate and distance that can be expected from a fiber-optic link.

Fluke Networks

Optical time domain reflectometers are used by VDV technicians to measure signal loss in fiber-optic cables.

Fiber-Optic Cable Loss Reports

John Q. Customer, Inc

Test Date: MM/DD/YY

Fiber-Optic Trunk #
Total Fiber Count: 72

Loss Budget: -2.5dB ⟵ ALLOWABLE SIGNAL LOSS

Buffer Tube	Fiber	Loss (dB)	Buffer Tube	Fiber	Loss (dB)	Buffer Tube	Fiber	Loss (dB)
Blue	1	-2.3	Orange	1	-2.0	Green	1	-2.0
	2	-2.2		2	-2.1		2	-2.1
	3	-2.0		3	-2.4		3	-2.3
	4	-1.8		4	-2.2		4	-2.4
	5	-2.2		5	-2.2		5	-2.0
	6	-2.5		6	-1.9		6	-1.9
	7	-2.1		7	-2.3		7	-1.8
	8	-2.2		8	-2.2		8	-2.3
	9	-1.7		9	-1.8		9	-2.2
	10	-1.9		10	-2.3		10	-2.1
	11	-2.3		11	-2.1		11	-2.2
	12	-2.8		12	-2.0		12	-2.4

FIBER BUNDLE

ATTENUATION TESTS ON INDIVIDUAL FIBERS

EXCEEDS ALLOWABLE SIGNAL LOSS

Buffer Tube	Fiber	Loss (dB)	Buffer Tube	Fiber	Loss (dB)	Buffer Tube	Fiber	Loss (dB)
Brown	1	-2.3	White	1	-2.0	Slate	1	-1.8
	2	-2.2		2	-2.1		2	-2.3
	3	-1.8		3	-2.3		3	-2.2
	4	-2.3		4	-2.4		4	-2.1
	5	-2.1		5	-2.0		5	-2.2
	6	-2.0		6	-1.9		6	-2.4
	7	-2.1		7	-2.0		7	-2.3
	8	-2.2		8	-2.1		8	-2.2
	9	-1.7		9	-2.4		9	-2.0
	10	-1.9		10	-2.2		10	-1.8
	11	-2.3		11	-2.2		11	-2.2
	12	-2.2		12	-1.9		12	-2.5

Tested by:___XXXX_____

(**NOTE:** Fiber 12 in the Blue Buffer Tube is not an acceptible result. This exceeds the calculated loss budget.)

Figure 3-11. A cable loss report is usually generated by a VDV technician to certify all of the fibers in a cable.

An OTDR sends a pulse of light along the length of a fiber and then detects the reflected light. By measuring the light intensity and graphing it against time, a technician can visualize how well the fiber is functioning.

Knowing the return trip time for a specific reflection allows the OTDR to calculate the length. The light intensity is also a linear relationship with signal loss. Less light means more signal loss. Therefore, although the OTDR is actually measuring reflected light intensity and time, it will also calculate and display signal loss and cable length. Reflections can come from imperfections or breaks in the fiber, in an end connector, or in splices. An OTDR is most useful as a troubleshooting tool and provides information such as the following:

- total fiber length
- attenuation of signal strength
- condition of any splices and connectors in the fiber

An OTDR is not as accurate as the optical attenuation test set for measuring loss because an OTDR uses reflected light instead of a direct measurement. However, it can be used to locate the break point in a nonfunctioning fiber link. Because it calculates distance, it can identify the point in the cable where the problem is located and convey that information through an OTDR report. **See Figure 3-12.**

Note: Often a test instrument OEM will produce a cable analyzer which can utilize different tester modules. This piece of equipment can then be used as a TSB-67 compliance tester, an optical attenuation test set, or an OTDR depending on how it is configured. These testers can generate reports and thereby eliminate the need to produce manual fiber cable loss reports.

Tech Tip

Fiber-optic cables are considered more secure than copper cables as they cannot be as easily spliced as copper cables to breach data.

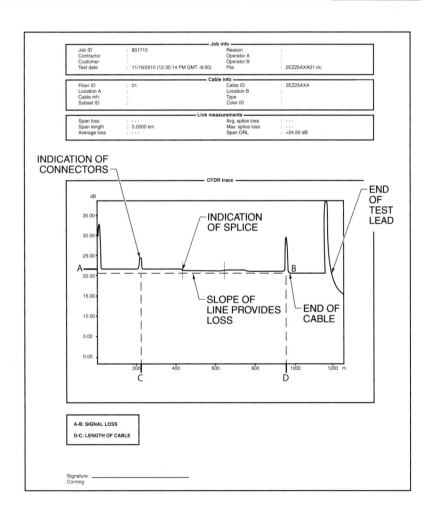

Optical Time Domain Reflectometer (OTDR) Reports

Figure 3-12. Because an OTDR measures distance, it can identify the point in a cable where a problem is located and convey that information through an OTDR report.

MOBILE ELECTRONICS

As technology has advanced, so has the use of common electronics in the VDV technician's job. Cellular phones have made individual users more accessible to each other. When testing any length of cable, the technicians must communicate frequently.

Often, a laptop computer is used to consolidate test data. In many instances, a computer is needed to access and program a piece of network equipment. Of course, many cell phones are now as sophisticated as computers and can serve many of the same functions.

A *global positioning system* (GPS) is a worldwide positioning system that uses signals from navigational satellites orbiting the Earth to determine the location and elevation of a GPS receiver. GPS receivers are commonly used on new construction sites, often through the use of survey equipment. Locations of manholes, handholes, and other buried junction points can be located through the use of GPS technology. **See Figure 3-13.** These access points may need to be located for testing, repairing, or rerouting buried VDV cabling.

Global Positioning Systems (GPSs)

GPS RECEIVER AS PART OF SURVEY EQUIPMENT

Figure 3-13. GPS receivers are commonly used on new construction sites, often through the use of survey equipment, to locate manholes, handholes, and other buried junction points.

Summary

VDV tools and test equipment are used for installation and testing. Most tools used by a VDV technician are hand tools such as hammers, wrenches, saws, crimpers, pliers, and wire strippers designed for specific use with VDV cables and devices.

Test instruments are used to troubleshoot faulty equipment and verify the integrity of equipment installations. VDV technicians must be knowledgeable of the types of tools and test equipment required for each job.

Chapter Review

1. List four types of hand tools that are primarily used in the electrical and VDV trades.

2. What type of copper cable is commonly used for VDV applications?

3. Explain how a punchdown tool works and its advantages.

4. Explain the differences between a fish tape and a glow rod.

5. What is another term for a wiremap tester?

6. List three parameters measured by a TSB-67 compliance tester.

7. What is a butt set?

8. Explain why a visual light source used to verify the continuity of a strand of fiber is red.

9. What is a cable loss report?

10. How are GPS receivers used when installing VDV systems?

Chapter Activity VDV Tool Identification

Identify each tool shown. Write each letter in the appropriate blank.

_____ **1.** adjustable wrench   _____ **6.** fish tape

_____ **2.** butt set   _____ **7.** glow rod

_____ **3.** cable cutters   _____ **8.** hacksaw

_____ **4.** drywall saw   _____ **9.** hammer

_____ **5.** electrician's scissors   _____ **10.** level

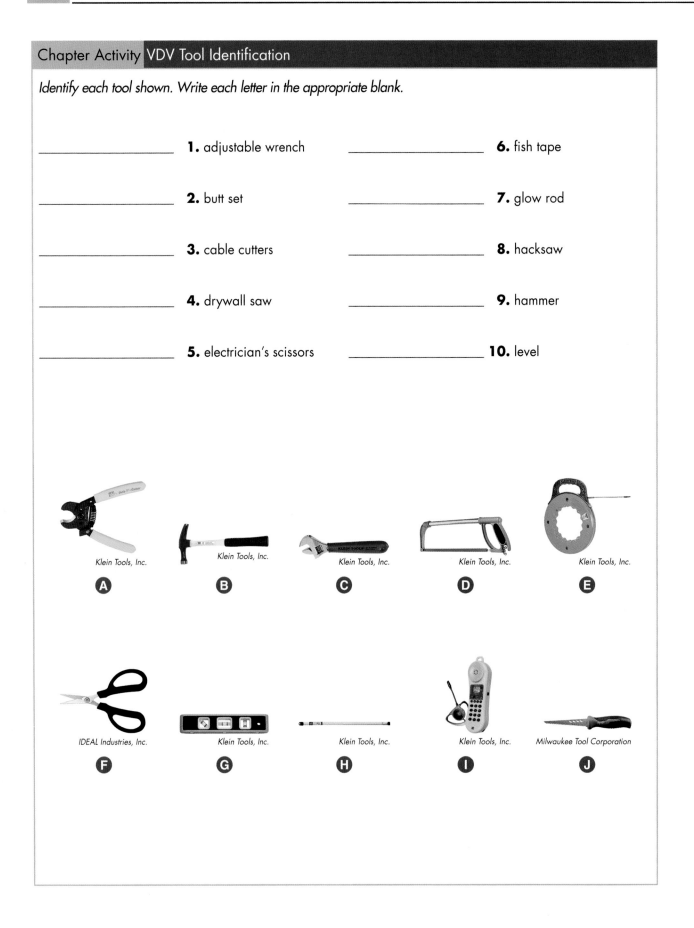

Klein Tools, Inc. **A**   *Klein Tools, Inc.* **B**   *Klein Tools, Inc.* **C**   *Klein Tools, Inc.* **D**   *Klein Tools, Inc.* **E**

IDEAL Industries, Inc. **F**   *Klein Tools, Inc.* **G**   *Klein Tools, Inc.* **H**   *Klein Tools, Inc.* **I**   *Milwaukee Tool Corporation* **J**

Chapter Activity VDV Tool Identification

Identify each tool shown. Write each letter in the appropriate blank.

_____ **11.** long-nose pliers

_____ **12.** punchdown tool (110)

_____ **13.** punchdown tool (66)

_____ **14.** tape measure

_____ **15.** tone generator

_____ **16.** tongue-and-groove pliers

_____ **17.** TSB-67 compliance tester

_____ **18.** utility knife

_____ **19.** VDV cable stripper

_____ **20.** wire map tester

_____ **21.** wire stripper

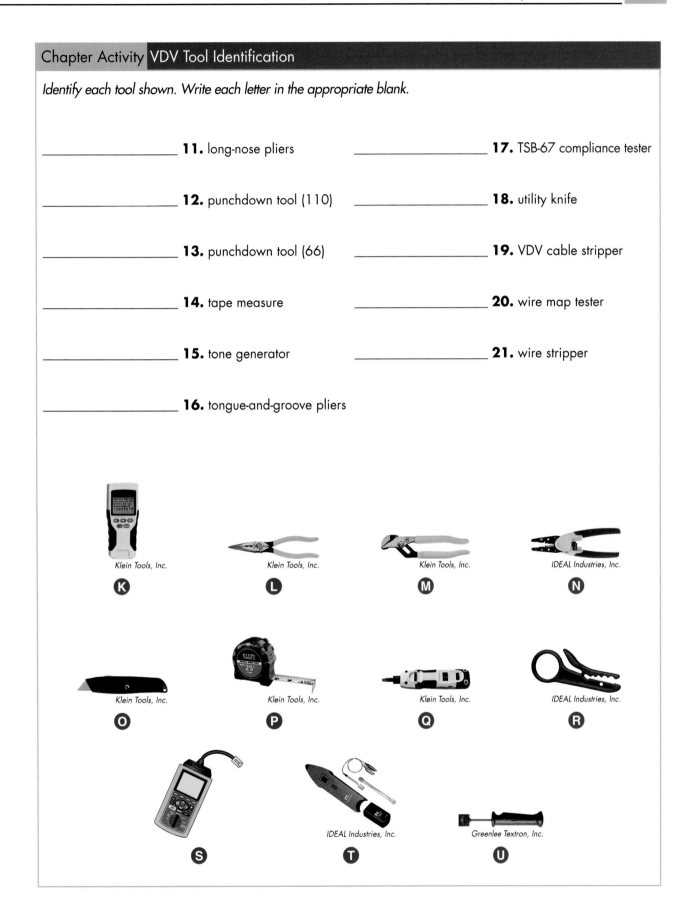

Klein Tools, Inc.
K

Klein Tools, Inc.
L

Klein Tools, Inc.
M

IDEAL Industries, Inc.
N

Klein Tools, Inc.
O

Klein Tools, Inc.
P

Klein Tools, Inc.
Q

IDEAL Industries, Inc.
R

S

IDEAL Industries, Inc.
T

Greenlee Textron, Inc.
U

Chapter 4

Copper Structured Cabling Systems

VDV systems are installed with cabling and associated components to form a structured cabling system. While a structured cabling system for a network can be either a copper system, a fiber-optic system, or both, most networks use copper systems. The copper cables can be either unshielded or, to reduce electromagnetic interference, shielded twisted pair. All cables and related devices must be installed in compliance with the NEC® and ANSI/TIA standards. Cabling is designed to meet specific application needs and is categorized accordingly. When installed as a complete system, many manufacturers provide a substantial warranty.

OBJECTIVES

- Describe how cables in copper structured cabling systems have evolved over the years.
- Explain the difference between twisted-pair cables and shielded-pair cables.
- Describe the differences between horizontal and backbone cabling.
- Identify how to apply NEC® ratings and ANSI/TIA requirements.

Digital Resources
ATPeResources.com/QuickLinks
Access Code 838502

LOW-VOLTAGE CABLE EVOLUTION

Copper wiring has been the standard used for communications systems since before the American inventor Alexander Graham Bell invented the telephone in 1876. Prior to the telephone, communication over a distance was often performed with a telegraph. The telegraph was a communication device that transmitted messages with electrical circuits of copper wire. The telegraph had only one switch, or key. The key was either closed (pressed down to close the circuit) or open (left up to open the circuit). **See Figure 4-1.** A message was sent by repeatedly opening and closing the key to form dots and dashes. A dot was a momentary closure of the key, and a dash was a slightly longer closure. A systematic code using these dots and dashes, developed by American scientist Samuel Morse in 1844, was the standard used for communication over a telegraph system. Patterns of dots and dashes were used to represent digits and letters, and these digits and letters were used to form words and sentences. It was a slow and tedious process, but messages could be transmitted over significant distances.

Telegraph systems used copper wire for 30 years prior to Bell's invention. It was Bell's attempt to create a harmonic telegraph, or one that could send multiple signals simultaneously over a single wire, which led to the invention of the telephone. The invention completely changed how messages were delivered. It allowed a person to simply speak into a device, and the message would be carried over a pair of copper wires as an electric signal and reconstructed into sound at the receiving end. Every active conversation required a pair of copper wires. Large cables containing many pairs of copper wires, or many circuits, were used to allow for simultaneous calls.

The invention of the telephone also resulted in other changes. Morse code gave way to the American Standard Code for Information Interchange or ASCII code. The first edition of ASCII code was published in 1963 by the American Standards Association (ASA). It was originally published in English and contained only limited characters. It has been revised several times and superseded by Windows-1258 and UTF-8, which are sometimes referred to as extended ASCII.

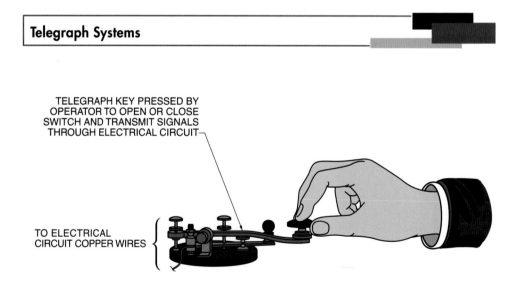

Telegraph Systems

TELEGRAPH KEY PRESSED BY OPERATOR TO OPEN OR CLOSE SWITCH AND TRANSMIT SIGNALS THROUGH ELECTRICAL CIRCUIT

TO ELECTRICAL CIRCUIT COPPER WIRES

Figure 4-1. With telegraphs, a single switch, or key, was either closed (pressed down to close the circuit) or open (left up to open the circuit).

The key, however, is that a binary code, or a code that uses only two states (on or off, zero or one, a short key dot or a long key dash), can be used to transmit coherent data rapidly. The dots and dashes used in the telegraph system were replaced with ones and zeros. A machine makes the conversion possible, and it is much faster than a person entering the code manually. Also, the ones and zeros can now be used to digitally represent pictures. The ones and zeros are referred to as bits of data. However, this system requires many bits in a short amount of time. **See Figure 4-2.**

ASCII CODE

Ctrl	Dec	Hex	Char	Code	Dec	Hex	Char	Dec	Hex	Char	Dec	Hex	Char
^@	0	00		NUL	32	20		64	40	@	96	60	`
^A	1	01		SOH	33	21	!	65	41	A	97	61	a
^B	2	02		STX	34	22	"	66	42	B	98	62	b
^C	3	03		ETX	35	23	#	67	43	C	99	63	c
^D	4	04		EOT	36	24	$	68	44	D	100	64	d
^E	5	05		ENQ	37	25	%	69	45	E	101	65	e
^F	6	06		ACK	38	26	&	70	46	F	102	66	f
^G	7	07		BEL	39	27	'	71	47	G	103	67	g
^H	8	08		BS	40	28	(	72	48	H	104	68	h
^I	9	09		HT	41	29	)	73	49	I	105	69	i
^J	10	0A		LF	42	2A	*	74	4A	J	106	6A	j
^K	11	0B		VT	43	2B	+	75	4B	K	107	6B	k
^L	12	0C		FF	44	2C	,	76	4C	L	108	6C	l
^M	13	0D		CR	45	2D	-	77	4D	M	109	6D	m
^N	14	0E		SO	46	2E	.	78	4E	N	110	6E	n
^O	15	0F		SI	47	2F	/	79	4F	O	111	6F	o
^P	16	10		DLE	48	30	0	80	50	P	112	70	p
^Q	17	11		DC1	49	31	1	81	51	Q	113	71	q
^R	18	12		DC2	50	32	2	82	52	R	114	72	r
^S	19	13		DC3	51	33	3	83	53	S	115	73	s
^T	20	14		DC4	52	34	4	84	54	T	116	74	t
^U	21	15		NAK	53	35	5	85	55	U	117	75	u
^V	22	16		SYN	54	36	6	86	56	V	118	76	v
^W	23	17		ETB	55	37	7	87	57	W	119	77	w
^X	24	18		CAN	56	38	8	88	58	X	120	78	x
^Y	25	19		EM	57	39	9	89	59	Y	121	79	y
^Z	26	1A		SUB	58	3A	:	90	5A	Z	122	7A	z
^[	27	1B		ESC	59	3B	;	91	5B	[	123	7B	{
^\	28	1C		FS	60	3C	<	92	5C	\	124	7C	|
^]	29	1D		GS	61	3D	=	93	5D	]	125	7D	}
^^	30	1E	▲	RS	62	3E	<	94	5E	^	126	7E	~
^-	31	1F	▼	US	63	3F	?	95	5F	—	127	7F	◆*

Using ASCII

Example: Send the word "Wire"

	Hex		Binary Bits	# of Bits
W	57	=	01010111	8
i	69	=	01101001	8
r	72	=	01110010	8
e	65	=	01100101	8

Total bits for one word=32

Figure 4-2. With ASCII, a binary code that uses only two states (zero and one) is used to transmit coherent data rapidly.

Twisted-pair cable is the most common type of cable used with VDV systems and equipment.

A disadvantage of high speeds across data cabling is noise, or interference, specifically electromagnetic interference. *Electromagnetic interference (EMI)* is interference in signal transmission or reception caused by the radiation from electric and magnetic fields. Any stray signal can interfere with the reception and proper decoding of a message. In order to send bits quickly enough to be useful, the technology of cable design required some changes to the original design.

TWISTED-PAIR CABLES

VDV circuits and components are connected using conductors. A *conductor* is a low-resistance metal that carries electricity to various parts of a circuit.

Conductors are available as wires or in groups of wires, such as cable and cord. A *wire* is an individual conductor. A *cable* is a group of two or more conductors within a common protective cover that is used to connect individual components. A *cord* is a group of two or more flexible conductors within a common protective cover that is used to deliver power to a load by means of a plug.

Most individual conductors are enclosed within an insulated jacket to isolate one conductor from another. The jackets also provide protection for the conductors, make them safer to handle, and help to meet code requirements. Some individual conductors, such as a ground wire, may be bare. The most common types of cables used in VDV applications are twisted-pair cables and shielded-pair cables.

A reliable method to improve signal transmission consists of simply twisting the copper wires of each pair around each other. A *twisted-pair cable* is a cable that consists of multiple pairs of insulated copper wires twisted around each other lengthwise to reduce interference from one wire to another. Each pair is twisted a different number of twists per inch to reduce the potential for signal interference between pairs. With modern VDV systems, twisted-pair cable is the most common type of transmission cable used.

The tightly twisted wire pair ensures that both wires experience exactly the same shape and volume of external noise. The same signal is sent through both wires, but the signal is inverted on one of them. At the receiving end of the wires, the inverted signal is reinverted back to its original orientation and added to the signal from the other wire. Since the noise on the second wire was also inverted, it is now an exact opposite of the noise on the first wire. Thus, the noises are cancelled out, leaving only the original signal. **See Figure 4-3.** For this reason, twisted-pair copper cable has become the standard cable used for local area networks and voice communications worldwide. A *local area network (LAN)* is a short distance VDV communications network used to link computers and peripheral devices under a standard control format. LANs are typically installed as part of a commercial building or multibuilding campus system.

Tech Tip

Copper or fiber-optic cables can be used for LANs, which must be able to carry a high bandwidth.

Twisted-Pair Cables and Noise Reduction

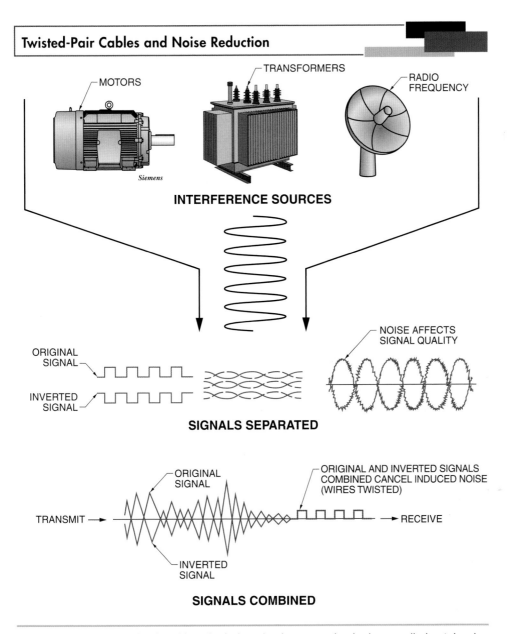

Figure 4-3. With twisted-pair cables, the induced noise on each wire is cancelled out, leaving only the original signal.

Ethernet is a signaling standard used in most LANs. It describes how groups of bits, or packets, are assembled and provides for protocol addressing and packet size (how many bits are transmitted at a time). Ethernet can be used with different types of cables. It was originally used with co-axial cable. However with modern systems, twisted-pair cable is almost always used for Ethernet. Token ring is another form of signaling protocol that uses a various cable types, but these systems are installed in a ring topology, leading from one computer to the next. However, Ethernet is installed in a star topology. Due to Ethernet's star topology and speed, token-ring systems are obsolete. There may still be some legacy token-ring systems in use. However, they are rare and difficult to maintain. *Topology* is the shape or arrangement of a network system. The most common topologies are bus, star, and token ring. **See Figure 4-4.**

Topology Patterns

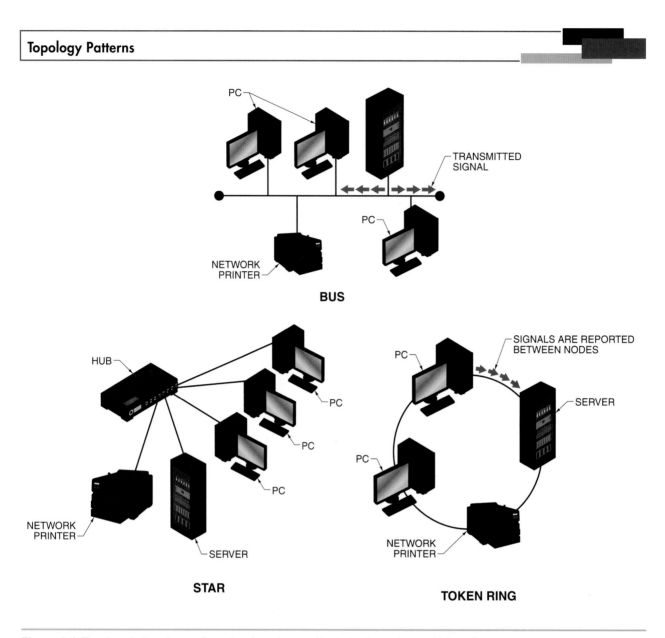

Figure 4-4. Topology is the shape of a network system and includes bus, star, and token ring.

Data communications outside of a building or a LAN can be delivered in a variety of ways, depending on the provider. Satellite systems use a dish mounted outdoors (on or near the building where the service is delivered) with a cable routed to a gateway or router in the building. Cable television service providers often use coaxial cable to deliver programming. Telephone service providers often use outside plant (OSP) paired cable or fiber-optic cable to deliver service.

Network systems have almost exclusively installed horizontal twisted-pair cables. Horizontal twisted-pair cables can be shielded or unshielded. Some cabling systems use both a twisted-pair design and shielding to get the best properties of both. These types of cables are simply referred to as shielded twisted-pair cable. A shielded twisted-pair (STP) cable is a cable that is comprised of wires that are separately insulated and twisted together in a spiral manner, with each pair wrapped within a metallic foil or braid to act as a shield from external sources of noise.

However, for reasons of cost and cable space requirements, unshielded twisted-pair cable is more commonly used. An unshielded twisted-pair (UTP) cable is comprised of wires that are separately insulated and twisted together in a spiral manner without metallic foil shields. The wires in a UTP cable are twisted around each other at regular intervals in order to confine the electromagnetic field within the wires. Higher category cables are often twisted more than twelve times per foot. By doing this, signal strength is maximized over distance, and interference between adjacent pairs in multipair cable is minimized. **See Figure 4-5.**

Additional data transmission cable types that may be seen by a VDV technician are pairs in metal foil (PiMF), unshielded foil-screened twisted pairs (U/FTP), and foil-shield unscreened twisted pairs (F/UTP). These cable types are not covered by ANSI/TIA standards. However, a VDV technician could see one or more of them in the field. An end user or specifying designer may use them in another installation.

PiMF cables are also known as double-shielded twisted-pair cables or shielded-shielded twisted-pair cables. PiMF cables are cables with multiple twisted pairs contained in the same cable jacket, with each twisted pair separately shielded by metallic foil and the entire grouping then shielded by another metallic foil layer.

U/FTP cables have an overall braid with foil-screened twisted pairs. Each individual pair is wrapped within a foil screen. This type of shielded cable is commonly used in 10GBase-T applications (twisted-pair systems that operate at 10 megabits per second, or 10 Mbps).

F/UTP cables are also known as foil twisted-pair cables or FTP cables. F/UTP cables are also commonly used in 10GBase-T applications. This type of cable is a standard four-pair cable that has a layer of foil under the outer sheath which covers all four pairs. It does not, however, have a screen or shielding over the individual pairs within the cable. This cable type is excellent for shielding out interference from other cables, but it does not protect the pairs from interference from adjacent pairs in the same cable.

Twisted-Pair Cables

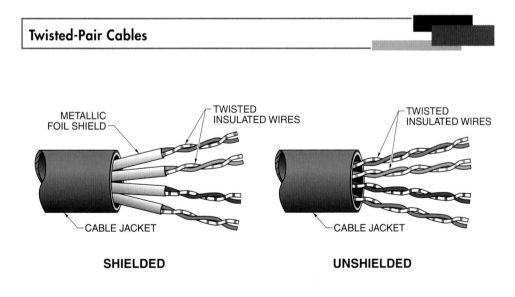

SHIELDED   **UNSHIELDED**

Figure 4-5. Twisted-pair cables can be shielded or unshielded. However, for reasons of cost and cable space requirements, unshielded twisted-pair, or UTP, cable is more commonly used.

UTP cables have become the standard for LANs in the United States. They are cost-effective, lightweight, and easy to terminate. The cables vary, however, in their data-carrying capacities depending on physical characteristics, such as how tightly twisted the wires are. Shielding, pair separation, and wire diameter may also have an impact on cable performance. **See Appendix.**

For this reason, a system of classifying twisted-pair cables is used. Categories of communication cable performance were developed for use as a sales tool by Anixter, Inc., a leading electrical and VDV equipment distributor, in the 1990s. Anixter originally defined three levels of cable as Level 1, Level 2, and Level 3.

Level 1 voice cable was often referred to as plain old telephone system (POTS) cable. It did not perform as well as twisted-pair cable, but it performed well enough for voice communication applications.

Level 2 ISDN and low-speed data cable is twisted-pair cable and has the capacity for up to 4 Mbps. This was the standard speed used for token-ring LANs promoted by IBM in the 1980s and 1990s. There are no shielded options available with Level 2 cabling.

Level 3 LAN and medium-speed data cable was designed for 10Base-T networks, which are networks that are 10 Mbps, nonmodulated over twisted pair. (The "T" represents "twisted pair.") Level 3 LAN and medium-speed data cabling was the original Ethernet twisted-pair standard and far exceeded token-ring LANs in application use. Levels 1 and 2 are sometimes referred to as Cat 1 and Cat 2, even though they do not fall under a Cat rating.

Cable Categories

The cable classification system was developed by ANSI/EIA/TIA (now TIA) and quickly adopted by many industries. This helped designers and installers understand what was necessary for computer network operations. With the increase in network speeds to 100 Mbps, TIA introduced the TIA-568 standard, now known simply as ANSI/TIA-568, which identifies cables according to categories, or Cat ratings. Categories are also defined in Standard ISO/IEC 11801.

Cat ratings define cables according to their data-carrying capacity. **See Figure 4-6.** When specifying a cable on a purchase order to a distributor or on a print, both NEC® and Cat ratings are required. For example, "plenum-rated Cat 6" is a common cable designation that would be seen on purchase orders or prints.

Color Codes

Cable jackets (outer coverings) are available in a selection of several different colors. The color of a cable jacket can be used to identify the cable later. For example, it is common to install one white cable for voice transmission and one blue cable for data transmission for each workstation in a work area. Cables with red jackets are usually avoided for voice or data installations because they are commonly used for fire-alarm system installations or backbone cabling. *Note*: Since fiber-optic system cables have orange-, yellow-, and teal-colored jackets, those colors are avoided for copper cables. White- and blue-colored cable jackets are the most commonly used jackets with copper cables.

However, the colors of the insulation on the wire pairs inside a cable are standard colors. In a four-pair cable, the color of the insulation on the wire pairs are blue and white or white-striped; orange and white or white-striped; green and white or white-striped; and brown and white or white-striped. *Note*: The striping pattern on white insulation can be vertical or horizontal. A standard color-code system identifies each pair of conductors in a cable. **See Figure 4-7.** These colors match the colors of the connectors to assure proper connections at both ends of the cable. Additional standard wire color codes are used for cables that contain up to 600 pairs and large cables that contain up to 4200 pairs are also available. The tip (positive) and ring (negative) conductors are also identified in conductor color-code charts.

VDV Cable Category Ratings

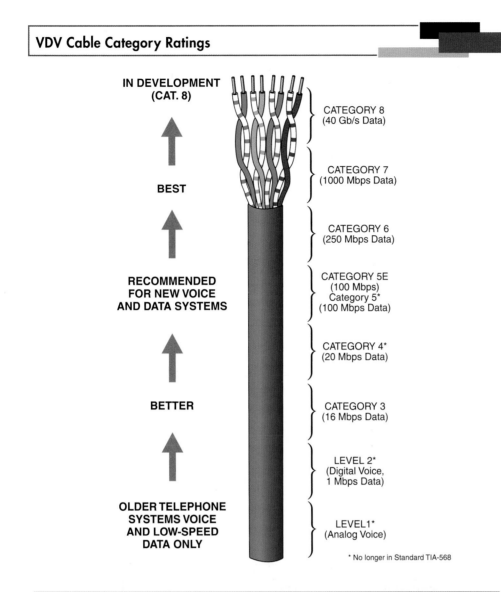

IN DEVELOPMENT
(CAT. 8)

BEST

RECOMMENDED
FOR NEW VOICE
AND DATA SYSTEMS

BETTER

OLDER TELEPHONE
SYSTEMS VOICE
AND LOW-SPEED
DATA ONLY

CATEGORY 8
(40 Gb/s Data)

CATEGORY 7
(1000 Mbps Data)

CATEGORY 6
(250 Mbps Data)

CATEGORY 5E
(100 Mbps)
Category 5*
(100 Mbps Data)

CATEGORY 4*
(20 Mbps Data)

CATEGORY 3
(16 Mbps Data)

LEVEL 2*
(Digital Voice,
1 Mbps Data)

LEVEL1*
(Analog Voice)

* No longer in Standard TIA-568

VDV Cable Classification System	
TIA–568a Categories	Comments
Level 1	Used with POTS cable installations; suitable for voice only applications; <1 MHz frequency capacity.
Level 2	Used with IBM token ring cable installations; 4 MHz frequency capacity; 1 Mbps data.
CAT 3	Used with 10Base-T cable installations; Original Ethernet; 16 MHz frequency capacity; 16 Mbps data.
CAT 4	Used with enhanced token ring instalations; 20 MHz frequency capacity; 20 Mbps data.
CAT 5	Originally described as "the last cable you'll ever need"; 100 MHz frequency capacity; 100 Mbps data.
CAT 5E	Enhanced Cat 5 or Cat 5e; eventually replaced Cat 5 entirely; 100 MHz frequency capacity at distances up 328′ (100 m); 100 Mbps data.
CAT 6	Used with gigabit Ethernet, 1000Base-T network cable installations; 250 MHz frequency capacity; twice the bandwidth capacity of Cat 5e cable; 250 Mbps data.
CAT 7	Used with 10 gigabit Ethernet cable installations; 600 MHz frequency capacity; mainly used in Europe; 1000 Mbps data.
CAT 8	In development; 40 Gb/s frequency capacity; 40 Gb/s data.

Figure 4-6. Levels of VDV cable performance were developed by ANSI/TIA/EIA.

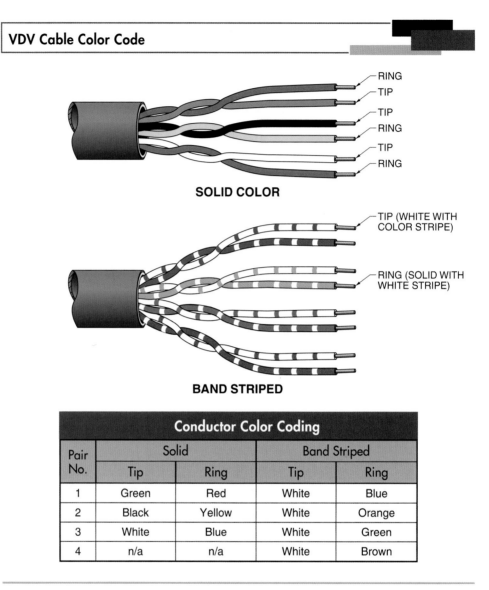

VDV Cable Color Code

SOLID COLOR

BAND STRIPED

Conductor Color Coding				
Pair No.	Solid		Band Striped	
	Tip	Ring	Tip	Ring
1	Green	Red	White	Blue
2	Black	Yellow	White	Orange
3	White	Blue	White	Green
4	n/a	n/a	White	Brown

Figure 4-7. A standard color-code system identifies each pair of conductors in a cable.

Connectors

The RJ-11 connector was originally developed by AT&T for its telephone systems. An RJ-11 connector is a modular jack and usually has connection pins for six wires, though it is typically only used with four wires. This is the most common type of telephone jack in the world. RJ-11 male plugs are used to connect telephones, modems, and fax machines to female RJ-11 jacks located on walls or floors.

The most common network connector in use today with copper VDV systems is the RJ-45 connector. **See Figure 4-8.**

Although larger in size, an RJ-45 connector resembles and operates similar to an RJ-11 connector. However, the RJ-45 connector has connection pins for all eight wires in a Cat 5, 5e, or 6 cable. The parallel manner in which the conductors are terminated in the connectors make it more difficult for the cable to resist noise interference. In spite of this, the RJ-45 connector is currently the standard connector used by most network cabling systems. Together with patch panels, jacks, and patch cables, these connectors and cables compose a horizontal UTP cabling system. **See Figure 4-9.**

Cable Connectivity

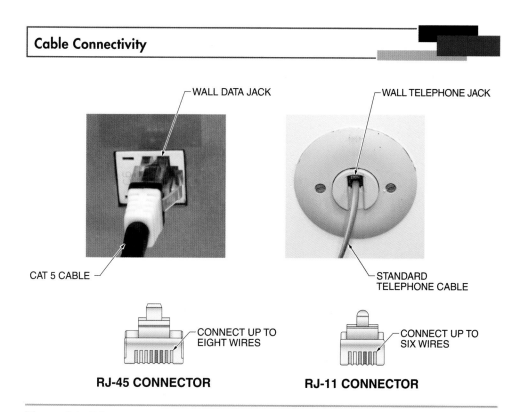

Figure 4-8. RJ-11 connectors were originally developed for telephone connections. RJ-45 connectors operate in a similar manner but are designed for Cat 5, 5e, and 6 connectivity.

Horizontal Cabling Systems

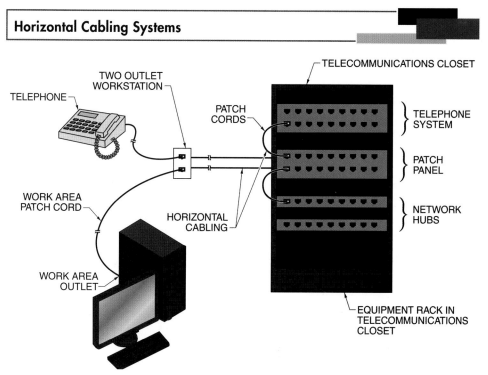

Figure 4-9. Patch panels, patch cords, connectors, and cable compose a horizontal UTP cabling system.

Backbone Twisted-Pair Cables

Backbone, or riser, twisted-pair cabling uses a design similar to twisted-pair cabling but typically uses higher pair-count cables. Horizontal cables typically provide a four-pair connection for each work area. However, backbone (and campus) cables use many more pairs to interconnect telecommunications rooms and buildings. Thus, backbone twisted-pair cables are heavier and larger in diameter than four-pair cables. **See Figure 4-10.** Cable rated for backbone installations is rated as "CMR" or "Communications Multipurpose Cable, Riser," which indicates that the cable is suitable for use in a backbone installation. This means it can be installed vertically between stories of a commercial building. (Backbone cable ratings are provided by the NEC®.) Backbone cables are available in multipair designs of 25, 100, 300, or 600 pairs as well as shielded versions. Shielded backbone cables are referred to as ARMM cables (armored).

Connectivity for multipair cables involves many types of terminations, but the most common termination components are the 66-style wiring block and the 110-style wiring block. Both types can be used for terminations in groups of 25-wire pairs. The original telephone block was the 66 block. The more modern 110 block has a higher capacity (more wire pairs can be terminated per square inch of wall space) and lower noise characteristics. However, the 66 block is less expensive and still used by many telephone service providers. **See Figure 4-11.**

Backbone Cabling System

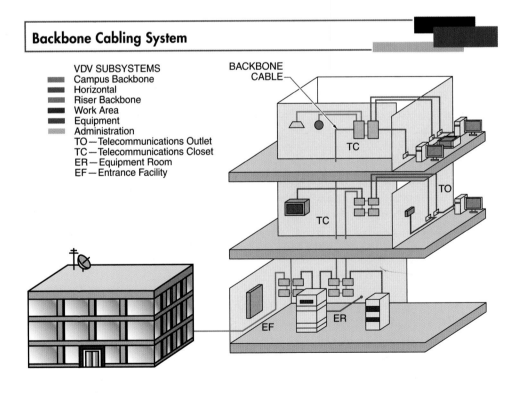

VDV SUBSYSTEMS
- Campus Backbone
- Horizontal
- Riser Backbone
- Work Area
- Equipment
- Administration
- TO—Telecommunications Outlet
- TC—Telecommunications Closet
- ER—Equipment Room
- EF—Entrance Facility

BACKBONE CABLE

Figure 4-10. Backbone twisted-pair cabling uses a design similar to twisted-pair cabling but typically uses higher pair-count cables.

Multipair Cable Terminations

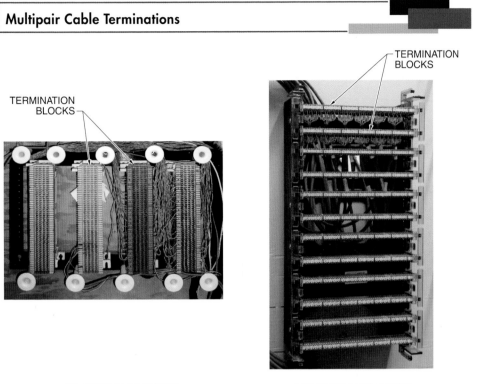

TERMINATION BLOCKS

TERMINATION BLOCKS

66-STYLE BLOCKS   **110-STYLE BLOCKS**

Figure 4-11. Connectivity for multipair cables most commonly involves the 66-style block and the 110-style block.

SHIELDED-PAIR CABLES

A *shielded-pair cable* is a cable with color-coded, insulated pairs of wires wrapped in sheaths, which are all wrapped within a metallic braid or foil to prevent the wires from picking up external signals or interference. It is important to shield copper cables from outside noise. The purpose of placing a metallic shield around the signal-carrying wires is to block unwanted noise. **See Figure 4-12.**

COAXIAL CABLES

The most common cable types used in VDV systems are coaxial cable and twisted-pair cable. Both types of cable are common but have differences in construction, costs, and applications. While coaxial cable was the standard cable used for VDV applications for many years, twisted-pair cable is the choice for new installations.

A *coaxial cable* is a low-voltage cable comprised of an insulated central conducting wire wrapped within an additional cylindrical shield. Coaxial cable is nearly impervious to external noise and is sturdy in its construction. The earliest data networks used a form of coaxial cable because of the excellent properties of the conducting wire. The foil, outer conductor, and braided shield in certain cables, form a shield around the inner conductor and add strength (and cost) to the cable. Coaxial cable is still used with some video applications such as cable television and some hardwired closed-circuit camera systems. The term "cable TV" arose from the coaxial cable commonly used to distribute the signals from cable-television service providers.

Shielded-Pair Cables

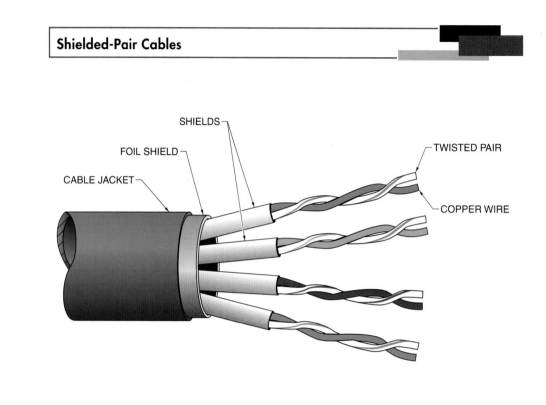

Figure 4-12. A shielded-pair cable is a cable with color-coded, insulated pairs of wires wrapped in sheaths, which are wrapped within a metallic braid or foil shield to prevent the wires from picking up external signals or interference.

When installing coaxial cables, connectors are also required. There are many different designs of coaxial cable connectors. F-type connectors are the most common because they are easy to install and fairly cost-effective. BNC connectors tend to be used for security applications. N-type connectors are only used in Ethernet applications. **See Figure 4-13.**

NATIONAL ELECTRICAL CODE® RATINGS

The National Electrical Code® (NEC®) provides ratings for VDV cables to indicate where they can be safely installed. Cable ratings for VDV systems are extensively covered in *NFPA 70, National Electrical Code,* Chapter 8—*Communications Systems.* Specifically, Table 800.179—*Communication Wires and Cables* identifies cables and lists them by acronym and application.

The NEC® rates cables by the degree to which they resist the spread of fire. Cable rated for a plenum environment (CMP) has the highest rating, while cable rated as riser communications cable (CMUC) has the lowest rating. A *plenum* is a duct or area connected to distribution ducts through which air moves. Air in a plenum may be maintained at a higher pressure than normal atmospheric pressure.

To avoid ordering a limited amount of CMUC cable for a project some installers may substitute riser or general-purpose cable for CMUC cable. Substituting a high-rated cable for a low-rated cable has a higher material cost. But if only a limited amount of low-rated cable is to be used within a project, it could actually be more cost effective to purchase only high-rated cable for the project. **See Figure 4-14.**

Coaxial Cables and Connectors

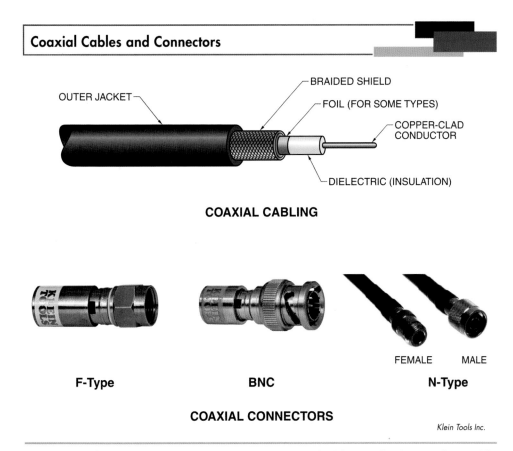

COAXIAL CABLING

F-Type   BNC   N-Type

FEMALE   MALE

COAXIAL CONNECTORS

Klein Tools Inc.

Figure 4-13. Coaxial cables and connectors are used with video applications such as cable television and hardwired camera systems.

NEC® Communication Cable Designations	
CMP	Communication cable for plenum-rated areas
CMR	Communication cable for riser-rated areas
CMG or CM	General-purpose cable (not plenum or riser)
CMX	Limited-use communications cable
CMUC	Riser communications cable

Figure 4-14. The NEC® lists cable types by acronym and application.

ANSI/TIA/EIA 568 REQUIREMENTS

When installing cabling for a LAN, almost all end users will need the cabling to conform to the requirements listed in the ANSI/TIA-568 standard. This standard is used for almost all project specifications, and it documents how a network should be cabled. All installers must be familiar with the standard and have a copy readily available for reference. Some of the guidelines of *Commercial Building Telecommunications Wiring Standard* ANSI/TIA-568 include the following:

- The maximum length for a horizontal cabling run is 100 m and includes the patch cords on both ends of the run. In the United States, this is usually referred to as the "300′ limit." This also provides the installer a safety margin on length, as 300′ is slightly less than 100 m. **See Appendix.**

- The maximum pulling tension for a Cat-rated cable is 25 lb (force).

- The minimum inside bend radius of a cable is four times the outside diameter under no-load conditions.

WARRANTIES AND PACKAGING

At one time, there were over 100 original equipment manufacturers (OEMs) producing Cat 5 and Cat 6 cables and related hardware. All OEM versions of this equipment were slightly different in construction, and all used different trade names for their systems. Some major points the OEMs had in common, however, were the warranties and packaging available for their products.

Warranties

When an installation company becomes certified in an OEM program, the company can receive extended warranties of installed equipment that are backed jointly by the contractor and the OEM. Generally, these are for 20 years to 25 years. Currently, there are fewer OEMs that offer complete solutions (every product needed to complete an installation) than there were in the past. However, there are still several manufacturers who produce the cable and related components used in today's network systems. **See Figure 4-15.** Because each OEM uses slightly different types of connector housings and termination layouts, their warranty programs usually require training on specific procedures of termination and testing.

> **Tech Tip**
> Cable packaged in boxes must only be removed by pulling the cable directly from the box to prevent entanglement.

Packaging

Most commonly, cable comes in reels. Horizontal cabling typically comes from an OEM in reels of 1000′ lengths. They are also available in 3000′ lengths. Most runs of cable are longer than 50′ or 60′. When installing cable in the field, there is usually some amount left over on a reel which is too short to be of use. As a standard practice, if there is less than 50′ remaining on a reel at the conclusion of a job, the remainder is simply discarded. Using larger reels means that the unusable portion of cable at the end of a reel can be cut by a third, resulting in less wasted material. However, a 3000′ reel of a specific cable type is, of course, three times the weight of a 1000′ reel of the same cable, which makes shipping and handling more expensive. Twisted-pair and coaxial cabling is also usually packaged on reels but ordered to meet specific length requirements. Unshielded twisted-pair cable is typically shipped by an OEM in cable spool boxes (quick pull boxes) in common lengths of 250′, 500′, and 1000′.

| Communication Cable, Connectivity, and Related Component OEMs ||
Company Name	Products
Belden	Complete solutions*
Berk-Tek Leviton Technologies	Copper and fiber-optic cables and connectivity
CommScope	Complete solutions*
Corning	Complete fiber-optic solutions and copper network interface devices
General Cable	Copper and fiber-optic cables
Panduit	Copper and fiber-optic connectivity
Siemon	Complete solutions*
Superior Essex	Copper and fiber-optic cables

* Connectors, cable, cabinets, racks, software, routers, PoE, fiber-optic, copper

Figure 4-15. There are several manufacturers that produce the cable and related components used in today's network systems.

Summary

Copper cabling has changed as technology has advanced. In today's buildings, Cat 5e, Cat 6, and coaxial cables are the three most commonly used cable types. An installer must be familiar with these cables and their connectivity.

In addition to NEC® and ANSI/TIA standards, both the physical environment and a cable's purpose must be considered when selecting cable types. Details such as jacket color can be important for future identification of cables and management of a structured cabling system. Most OEMs provide warranties and training for cable installers. Cable packaging is dependent on the cable type required and the size of the job.

Chapter Review

1. What are the four colors used on the insulation of the separate wire pairs in a four-pair copper VDV cable?

2. What are the two design methods used to minimize noise interference in VDV cables?

3. According to ANSI/TIA-568, what is the maximum installed length of a horizontal Cat 6 cable?

4. Is it acceptable to install a plenum-rated cable in a riser-rated space?

5. What are the most common colors used on the outer covering, or jacket, of a copper cable?

6. Explain the difference between a cable and a cord.

7. What change was made in cable design to improve the speed of signal transmission?

8. How is shielded cable constructed?

9. Explain the initial development of Cat ratings.

Chapter Review

10. What is the length of time of a typical extended warranty for installed VDV equipment?

11. What is the most common form of packaging for VDV cable?

12. According to ANSI/TIA-568, what is the maximum pulling tension for a Cat-rated cable?

13. According to ANSI/TIA-568, what is the minimum inside bend radius of a VDV cable?

14. According to the NEC®, what is CMUC cable?

15. According to the NEC®, what is CMP cable designated for?

Chapter Activity VDV Acronyms

Provide the NEC® acronym for the following types of communication cable.

_____ **1.** Limited-use communications cable

_____ **2.** Communication cables for riser-rated areas

_____ **3.** Riser communications cable

_____ **4.** Communications cables for plenum-rated areas

_____ **5.** General-purpose cables (not plenum or riser)Provide the NEC® acronym for the following types of communication cable.

Chapter 5

Fiber-Optic Cabling Systems

While many VDV systems consist of copper-structured cabling and related components, others are installed with fiber-optic cables, connectors, and related components. Although copper cables are less expensive and more common, fiber-optic cables offer several advantages including less space needed for installation, higher security, a resistance to electromagnetic interference, and an ability to transmit greater amounts of data at faster speeds. Because of this, VDV technicians must be familiar with fiber-optic hardware, installation techniques, test methods, and applications.

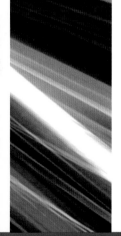

OBJECTIVES

- Compare light transmission and electrical transmission.
- Describe fiber-optic digital communications.
- Explain light propagation and measurement.
- List the different types of fiber-optic cables and related connectors.
- Describe how fiber-optic cables are constructed.
- Describe OEM fiber-optic equipment warranties.

Digital Resources

ATPeResources.com/QuickLinks
Access Code 838502

FIBER-OPTIC SYSTEMS VS COPPER-BASED SYSTEMS

In copper-based communications systems, voltage levels are used to transmit signals. In fiber-optic communications systems, pulses of light are used to transmit signals. The physical characteristics of fiber-optic systems are considerably different from copper-based systems. The transmission material in a fiber-optic cable is composed of glass rather than copper, and the delivered signals are optical rather than electrical.

Fiber-Optic System Advantages

Fiber-optic cabling systems have become standard for applications that require high-speed data transmission, increased data-carrying capacity, lightweight cabling, the ability to transmit data over long distances, and/or immunity to electromagnetic interference or detection. Fiber-optic systems also offer greater security than copper systems as they are difficult to tap into or splice by unknown sources. The advantages of fiber-optic systems, however, come with some financial and technical parameters that are different than copper systems, which must be considered.

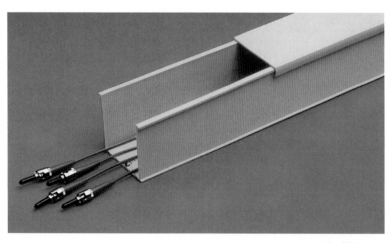

Panduit Corp.

Because fiber-optic cables have a lightweight design, smaller cable trays can be used to route them throughout an installation.

Overall System Cost

A fiber-optic system is generally more expensive than a twisted-pair copper system. Although Category 6A, 10 Gb cables can be nearly as costly as fiber-optic cables with a low number of fibers, the system electronics, patch panels, cords, terminations, and test methods are more costly for fiber-optic systems. Due to advances in technology and increased demand, terminations of fiber connectors have become less complex, and the equipment used to install fiber-optic systems has become less costly. While these changes in technology have made fiber-optic systems more affordable than they have been in the past, they are still considerably more expensive than twisted-pair copper-based cabling systems.

Technical Complexity

Optical systems, including fiber optics, involve refraction, light generation, and signal recognition. *Refraction* is the reflection of light within a glass strand caused by the differing densities of glass used in the core and cladding. With a glass strand, the core has higher density than the cladding. So when a ray of light meets the boundary between the core and the cladding, it is reflected back into the core. The difference in the index of refraction is how a fiber-optic cable works and is sometimes referred to as total internal reflection. Fiber-optic systems are more complex than electrical wiring systems, and they have less in common with electrical wiring than twisted-pair systems.

FIBER-OPTIC CABLE CONSTRUCTION

A strand of fiber-optic glass is composed of a core, a cladding, and a coating. The *core* is the smallest section of a glass fiber strand and is used to carry light waves. It is composed of glass made from silicon dioxide (SiO_2) and has the fewest possible impurities that could interfere with light transmission. In the past, fiber-optic cable has been manufactured in various core sizes. In practice, however, there are three core sizes that have become the standards for nearly all applications. These core diameters are 8.3 μm (single-mode), 50 μm, and 62.5 μm (multimode), which are smaller in diameter than a human hair. This standardization of core sizes has made it possible for interoperability between equipment, connecting hardware, patch cords, and cable from different OEMs. Devices made for a specific type of light source by one OEM can be intercharged with any other OEM's cable.

The *cladding* is a layer of glass fused to a glass fiber core to aid in performance of light transmission and act as protection for the core. It is denser than the core and causes light to reflect back into it. The typical cladding diameter used for VDV applications is 125 μm. The *coating* is a layer of protective material applied to a fiber-optic glass strand. The coating layer protects the cladding from moisture, scrapes, shocks, and other events that could cause permanent damage to the strand. OEMs use standard color codes with coating layers to make it possible to identify individual strands.

Tech Tip

Plastic optical fiber (POF) is used for light propagation for short, low-bandwidth acoustical applications. Ideal POF applications include electronics in underwater areas, such as controlling lights in a pool. Acoustical applications for POF include short, low-bandwidth connections, such as the interconnection of a sound amplifier to a television. POF is inexpensive and sturdy but impractical for high-speed, long-distance data transmission.

A *buffer* is a fiber-optic cable layer that provides protection for optical fiber within a cable. The buffer protects the fibers from bumps, scrapes, stresses, and outside forces that could cause permanent damage. A *strength member* is a fiber-optic

cable component that runs the length of the cable and is used to increase cable tensile (pulling) strength. Common materials used for strength members in fiber-optic cable include aramid yarn and fiberglass or steel rods. A *jacket* is the outer protective layer of a fiber-optic cable. The jacket is typically composed of a flexible polymeric material such as PVC and is used to protect the internal cable components from oil, dirt, animal damage, water, ice, accidental contact with other equipment, rough handling, and exposure to chemicals. **See Figure 5-1.**

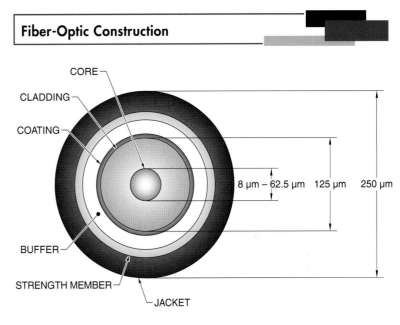

Fiber-Optic Construction

Figure 5-1. Fiber-optic cable is composed of an extremely thin strand of pure glass covered with strength members and insulation.

A *cable sheath* is an outer layer used to provide protection for a cable with multiple fiber strands. A *rip cord* is a thread installed along a cable jacket or sheath that allows the cable jacket or sheath to be split for access to individual fibers. A rip cord reduces the risk of damaging fibers with cutting tools.

Note: Aramid yarn is extremely tough and cannot be cut with ordinary cutting tools. A tool known as aramid scissors, or a Kevlar® cutter, must be used to cut aramid yarn.

Light Modes

A *light mode* is a physical path taken by a ray of light. When a light source is introduced into the transmission end of a fiber, many modes, or directions, of light are transmitted into the fiber. Those modes that are transmitted into the cladding are dissipated and lost. Those that are coupled into the core travel along the low-loss glass and arrive at the receiving end where they are interpreted as 1s or 0s and decoded by a photoelectric sensor back to the original message. One way to increase the speed of a fiber-optic cable is to manufacture glass strands with the core diameter small enough so that only one mode of light can travel down the core. Fiber is typically classified as multimode or single mode. **See Figure 5-2.**

Multimode Fiber

A *multimode fiber* is a glass fiber in which the core diameter is large enough to allow multiple light rays to travel through it. Multimode fiber-optic cable allows for less expensive light transmission and receiving equipment to be used, reducing the overall cost of the system. The most common configurations of multimode cable use core sizes of 62.5 µm and 50 µm. **See Figure 5-3.** There is a great amount of 62.5 µm cable in use in the United States. For this reason, there is a need to service, repair, and maintain it. However, most new installations use 50 µm cable. It is close in cost to the slightly larger core size of 62.5 µm, but it has less signal loss and higher bandwidth capacity. *Bandwidth* is the amount of data that can be sent through a given channel and is measured in bits per second (bps).

Single-Mode Fiber

A *single-mode fiber* is a glass fiber in which the core diameter is sized for only one light ray to travel through it. Cable made with single-mode fiber is actually less expensive to manufacture than multimode fiber cable. The typical core diameter of a single-mode fiber cable is between 8.3 µm and 11 µm. The disadvantage of single-mode cabling systems is the high cost of the light source.

A *light-emitting diode (LED)* is a semiconductor device that emits light when an electric current is passed through it. While multimode cable can be used with a relatively inexpensive light-emitting diode, single-mode cable requires a more expensive laser light source for proper operation. However, installations using single-mode cables can attain much higher transmission speeds over greater distances than multimode cables. Single-mode cables typically have a yellow-colored jacket. Although the electronics used for single-mode cable are more expensive, data centers use single-mode cable systems almost exclusively because of the increased data-carrying capacity. **See Figure 5-4.**

Light Modes

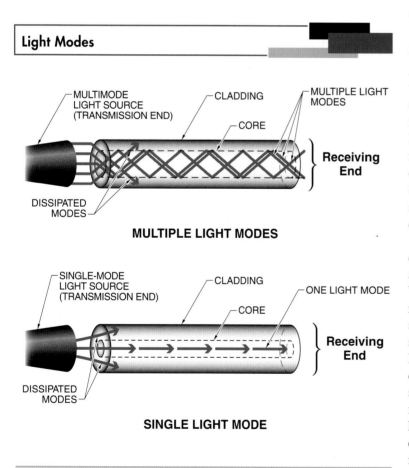

MULTIPLE LIGHT MODES

SINGLE LIGHT MODE

Figure 5-2. When a light source is introduced into the transmission end of a fiber, many modes, or directions, of light are coupled into the fiber with some traveling through the core to the receiving end.

Multimode Fiber Cables

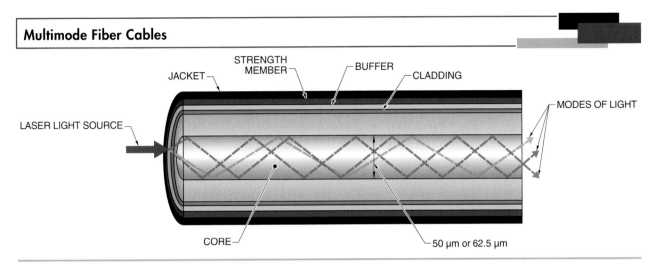

Figure 5-3. With a multimode fiber cable, the core diameter is large enough to allow multiple light rays from an LED to travel through it.

Single-Mode Fiber Cables

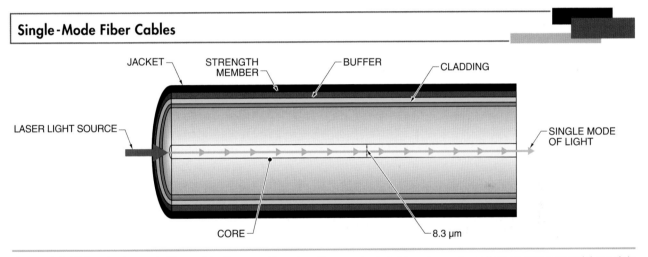

Figure 5-4. With a single-mode fiber cable, the core diameter allows a single light ray from a laser light source to travel through it.

FIBER-OPTIC PERFORMANCE FACTORS

When installing fiber-optic cable and related equipment, there are several factors that can negatively affect performance. The main factors that a VDV technician must be aware of are attenuation, modal dispersion, chromatic dispersion, acceptance angle, and numerical aperture.

Attenuation

As with copper systems, fiber-optic systems are affected by attenuation, or signal loss. With fiber-optic systems, attenuation is caused by the loss of light signals and is measured in decibels (dB) or decibels per meter (dB/m). Attenuation has a negative impact on transmission speeds as well as the usable distance of installed cable. Fiber-optic systems are typically more sensitive to attenuation than copper systems. Factors that cause attenuation include but are not limited to the following:

- contaminants on fiber ends
- an excessive gap between fiber ends
- improper terminations
- glass impurities
- exceeding bend radii
- excessive tensile stresses on cable during construction

Modal Dispersion

Modal dispersion is the slowing of bandwidth caused by the different rates of speed between modes of light through a fiber-optic strand. It is also referred to as differential mode delay (DMD). Because the paths of differing modes of light each have a different length, this causes the modes to spread out, and not all of them arrive at the opposite end of a fiber at the same time. Those modes that travel straight, or nearly straight, down the center of a fiber arrive sooner than those that have traveled a longer, reflecting path at the edges of the core.

Modal dispersion is a function of both distance and transmission speed. At a given distance, the maximum data speed that can be transmitted is the speed at which the pulses are still distinguishable from the previous and next pulse. For example, mode A is faster than mode B or mode C because its entrance angle is less severe and it has less distance to travel than the other modes. The difference in travel time between mode A and either mode B or C is the DMD. **See Figure 5-5.** Modal dispersion typically affects bandwidth in multimode cables. The higher the DMD in a fiber, the lower the bandwidth.

Multimode fiber-optic cables are typically installed in large commercial buildings or campuses.

Tech Tip

When fiber-optic systems were originally being developed, there was discussion among designers about what the optimum core size should be. The two sizes most often suggested were 50 μm and 75 μm. The 62.5 μm size was a compromise as it is exactly halfway between 50 μm and 75 μm. Later, as modal dispersion became an impediment to transmission speeds, 50 μm cable was reintroduced and is now more common than 62.5 μm for most applications.

Chromatic Dispersion

Chromatic dispersion is the slowing of bandwidth caused by the spread of a light pulse due to the varying refraction rates of the different colored wavelengths. Chromatic dispersion is a fiber-optic performance factor that mostly affects bandwidth with certain types of single-mode fibers, but it can affect multimode fibers as well. As a laser pulse travels the length of a fiber, the combined wavelengths spread apart until they arrive at the receiving end at different times. This will cause the laser light to be less sharp, wider, and lower in power. Both modal and chromatic dispersion can affect the bandwidth of multimode fiber. However, single-mode fiber bandwidth is typically only affected by chromatic dispersion. **See Figure 5-6.**

Acceptance Angle

An *acceptance angle* is the maximum angle at which light can enter a glass fiber core as an input and continue to reflect off the boundary layer between the core and the cladding. Acceptance angle typically affects multimode fibers. The greater the difference in acceptance angles between multiple signals, the greater the modal dispersion. A high acceptance angle can increase modal dispersion and have a negative effect on performance. An acceptance cone can illustrate the acceptance angle area for a multimode fiber. **See Figure 5-7.**

Modal Dispersion

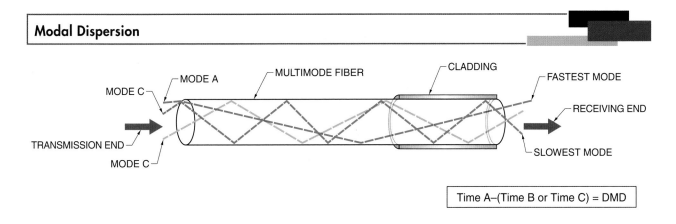

Time A–(Time B or Time C) = DMD

Figure 5-5. Modal dispersion, or differential mode delay (DMD), is the slowing of bandwidth caused by the different rates of speed between modes of light through a fiber-optic strand.

Chromatic Dispersion

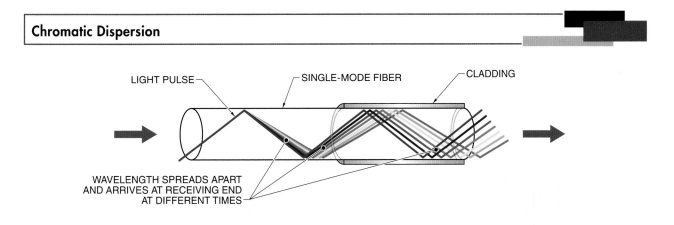

Figure 5-6. With chromatic dispersion, a light pulse travels the length of a fiber and spreads apart until it arrives at the receiving end at different times.

Fiber-Optic Acceptance Angles

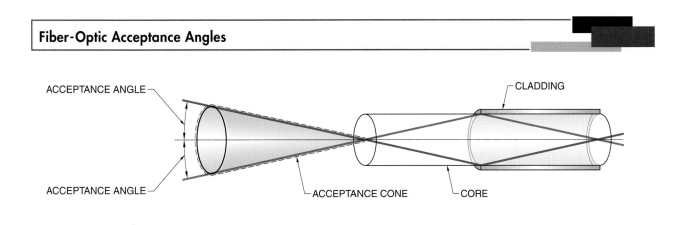

Figure 5-7. The greater the difference in acceptance angles between multiple signals, the greater the effect on modal dispersion.

Numerical Aperture

Numerical aperture (NA) is the light-accepting ability of a glass fiber. The NA of a fiber is determined by the refractive indexes of the fiber's core and cladding and is also used to determine the fiber's acceptance angle. A *refractive index* is the ratio of the velocity of light between the core and the cladding of an optical fiber. For example, a NA value of 0 indicates that the fiber can accept no light, while a value of 1 indicates that the fiber can accept all the light that it is exposed to. A higher NA value indicates that the fiber can accept light from a wider variety of acceptance angles. Multiple fiber typically has higher NA values than single-mode fibers.

NEC® FIBER-OPTIC CABLE DESIGNATIONS

As with electrical and low-voltage copper cables, the NEC® classifies a fiber-optic cable according to its construction and the physical environment in which it may be installed. A *conductive cable* is a cable that contains a metal component of some type, which would be subject to induced electrical current in the event of a lightning strike or physical contact with an energized electrical component. A *nonconductive cable* is a cable that does not have metal components. Therefore, it cannot conduct electricity under any circumstances.

Armored fiber-optic cable is a fiber-optic cable that utilizes an interlocking metallic armored design which eliminates the need to install rigid conduit to meet building codes. The metallic armor provides added protection for installation in high-traffic areas where security is required. It is ideal for industrial networking applications and is more flexible than corrugated steel armored cables. Armored fiber-optic cable is similar to electrical MC cable.

To indicate a cable as a fiber-optic cable, its designation begins with the letters "OF" for optical fiber. **See Figure 5-8.** The following letters in the designation indicate if metal exists in the cable and the environment in which

the cable may be installed:
- C–contains metal, (conductive)
- N–does not contain metal, (nonconductive)
- G–general purpose environment installations
- R–riser installations
- P–plenum installations
- LS-low-smoke environments

Tech Tip

In the telecommunications industry, glass fibers are also referred to as "tubes" or "light pipes."

FIBER-OPTIC CABLE TYPES

Fiber-optic cables are lightweight and offer the highest bandwidth and speeds possible compared to copper cables. Fiber optics also use smaller diameter cables, which are not affected by electromagnetic interference or radio frequency interference. Fiber-optic cables are used in backbone systems and sometimes at individual workstations, and the connectors must be protected from physical damage and moisture. To protect the fibers in cables, protective cladding and coatings are used along with strength members.

There are various types of fiber-optic cables available to meet various cable environments. Fiber-optic cables typically consist of more than one strand. In addition to standard fiber-optic cable, two other types of fiber-optic cable are the tight-buffered fiber-optic cable and the loose-tube fiber-optic cable. **See Figure 5-9.**

Tight vs Loose Cable

A *tight-buffered cable* is a fiber-optic cable that has the glass strand tightly attached to the buffer tube so that they move as a single construct. For indoor cable, fiber is placed in a buffer tube, which is tightly fitted over the fiber. However in most environments, extreme temperature conditions can pose a risk to this type of cable because the insulation, coating, and cladding can become damaged over time as individual fibers expand and contract.

NEC® Fiber-Optic Cable Designations	
OFN	Optical fiber nonconductive
OFC	Optical fiber conductive
OFNG or OFCG	General purpose
OFNR or OFCR	Riser-rated
OFNP or OFCP	Plenum-rated for use in air handling plenums
OFN-LS	Low-smoke density

Figure 5-8. The NEC® classifies a fiber-optic cable according to its construction and the physical environment in which it may be installed.

Fiber-Optic Cables

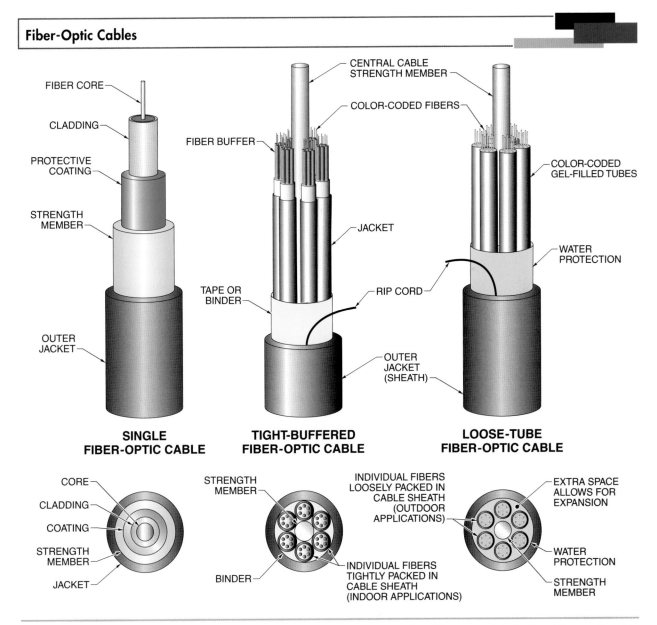

Figure 5-9. In addition to standard fiber-optic cable, two other types of fiber-optic cable are the tight-buffered fiber-optic cable and the loose-tube fiber-optic cable.

One method to make a fiber-optic cable suitable for hot or freezing environments is to utilize a loose-tube construction. The cable can then rely directly on this tube for strength and protection of the fiber. However in extreme temperature conditions, the glass fiber and the plastic buffer tube expand and contract at different rates. For this reason, the fiber can be placed in a larger buffer tube allowing slack in the fiber. For example, in extremely warm conditions, the buffer tube and fiber can expand at different rates without damage to the cable.

A *loose-tube cable* is a fiber-optic cable in which the buffer tube surrounding the glass strand is larger than the outer coating on the glass. The glass strand floats free inside the buffer tube and can expand and contract safely. Loose-tube fiber-optic cable has fiber that is not restricted and can move freely inside the housing. Although both types use air- or gel-filled tubes to allow fiber movement, tight-buffered fiber-optic cable typically uses gel-filled tubes. Loose-tube fiber-optic cable is stronger and is used in applications requiring long-distance cable runs.

Ribbon Fiber-Optic Cables

In order to achieve a high number of fibers in a small cable, ribbon fiber-optic cable, or ribbon fiber, was developed. Ribbon fiber gets its name from its appearance. Several strands (usually 12) are laid parallel. They are coated with a thin layer of plastic to keep them in this configuration, but they do not have any buffers. Several ribbons can then be layered and inserted in a cable for high fiber counts. Ribbon fiber is typically used where installation space and cable weight are a concern, such as in an existing conduit with little room for additional cabling. Ribbon cables can have fiber placed side by side in one layer or can be stacked as high as 22 ribbons in a loose-tube ribbon cable. **See Figure 5-10.**

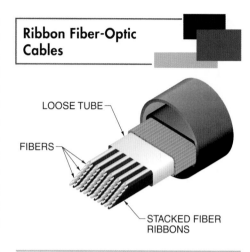

Ribbon Fiber-Optic Cables

LOOSE TUBE

FIBERS

STACKED FIBER RIBBONS

Figure 5-10. Ribbon fiber is typically used where installation space and cable weight are a concern.

Indoor Cables and Outdoor Cables

Traditional outdoor cables have an outer sheath made of polyethylene. Polyethylene has excellent tensile (pulling force) strength and is impervious to physical damage and penetration by water. However, due to its chemical properties, it melts easily and gives off toxic smoke when exposed to heat. Therefore, it cannot be used for installations indoors. When routing from outdoors to indoors, the outdoor cable must be terminated, and an indoor cable must be used instead. This process is expensive, introduces additional signal loss, and is time consuming. Also, outdoor cable is typically buried in the earth.

Some fiber-optic cables have alternate types of water blocking. Using a liquid fill or water-absorbing tape, the cables can then be installed in an outdoor environment but still have a sheath rated for indoor installation. This cable, sometimes called a "freedom" or indoor/outdoor cable, can be used to make the transition from outdoors to indoors without splicing to a different cable.

Cable Jacket and Sheath Colors

Although most outdoor cable jackets and sheaths are black, indoor fiber-optic cables vary in color but follow general guidelines.

> **Tech Tip**
>
> Ribbon fiber-optic cables are typically cost competitive in applications with fiber counts above 96.

An orange cable jacket or sheath generally indicates that the fiber within is 62.5 μm/ 125 μm multimode cable, which means that it has a core diameter of 62.5 μm and a cladding (outside) diameter of 125 μm. Multimode cable, 50 μm/125 μm, most often has a cable jacket or sheath that is aqua (blue-green).

Single-mode cable usually has a core diameter of 8.3 μm to 11 μm with an outside diameter of 125 μm and most often has a cable jacket or sheath that is yellow.

FIBER-OPTIC CABLE STRAND COUNTS

The most common number of fibers in a cable is 12 or a multiple of 12, such as 24, 36, or 48. A fiber-optic link requires one fiber for transmission and one fiber for reception. Therefore 6 links are provided by 12 strands. The top speed with this technology is 10 gigabits per second (10 Gbps). But data centers need faster speeds to meet data-processing requirements brought on by smart phones, the Internet, and media-streaming applications. So the current solution relies on parallel fiber optics.

With parallel fiber optics, it is possible to reach a standard speed of 40 Gbps. To achieve this speed, 4 glass strands are used to transmit and receive. All 8 strands are included in a single connector. Since most cables are currently manufactured with multiples of 12, 4 of the strands are unused. For this reason, some OEMs are starting to produce cables with multiples of 8 instead of 12.

FIBER-OPTIC CONNECTORS

Connectors for fiber-optic systems have changed dramatically over the last decade. A *connector* is a device that connects wires or fibers in a cable to equipment or other wires and fibers. With fiber-optic equipment, a connector is the physical means of terminating a glass fiber strand and holding it in perfect alignment with another glass fiber strand or the receiving hardware. Unlike copper wires, which only have to make physical contact with one another to pass a signal, optical fibers must be squarely cut (cleaved) on the ends and perfectly aligned. Fiber that is not perfectly cleaved and aligned can result in signal distortion, signal loss, or no signal at all. The connector (male) on one fiber-optic cable connects to a receptacle or adapter (female) on the opposite fiber-optic cable or port. The main techniques that have been used to attach connectors to a strand of glass are adhesive-type connectors, field-polish connectors, splice-type connectors, and prepolish splice-type connectors.

Fiber-optic connectors consist of a ferrule, a retaining nut or retainer, a backshell, a boot, and a dust cap. The ferrule (tip) precisely aligns the fiber in the connector. **See Figure 5-11.** When terminating or splicing a fiber-optic strand, core alignment is critical. Any misalignment of the cores will result in signal loss. Connectors and splices are designed to precisely align cores. Misalignment is usually the cause of failure in a fiber-optic cabling system. For best results, always follow OEM instructions provided with connectors. Most ferrules are ceramic. The retaining nut or retainer secures the connector to a receptacle or an adapter. The backshell is a rigid plastic component located behind the retaining nut and provides the main source of strength between the cable and connector. The boot, sometimes made from heat-shrink tubing, fits over the backshell, cable, and strength members and helps maintain the maximum bend radius that a cable can withstand.

A *bend radius* is the minimum radius that a cable can bend without causing damage. The dust cap protects the ferrule and glass fiber from the contamination and abrasion that can occur during ordinary handling of the connector. The dust cap is removed from the ferrule prior to installation.

Connector Performance

In addition to fiber-optic cables and light transmission properties, fiber-optic system performance is affected by connector performance. Fiber-optic connector performance is determined by taking measurements such as maximum signal loss per connector pair, average signal loss per connector pair, repeatability, and reflectance.

The maximum signal loss per connector pair is the maximum amount of signal loss that occurs in a pair of properly installed connectors and is measured in decibels (dB). The average signal loss per connector pair is the typical signal loss that occurs in a pair of properly installed connectors. Repeatability is the maximum change in signal loss per connector pair that can occur between signal measurements. Reflectance, also known as return loss or back reflection, occurs when connectors are improperly installed or terminated. Excessive reflectance can burn out a light transmitter.

Connector Styles

The styles of connectors used with fiber optics are SC, ST, FC, FDDI, and LC connectors. These connectors are available in simplex (one-fiber) or duplex (two-fiber) designs. Array connectors are used for multiple fiber cable connections. The connector style used depends on the application, the installation space available, and bandwidth requirements. For example, FDDI connectors are used with 100 Mbps LANs.

Fiber-Optic Connectors

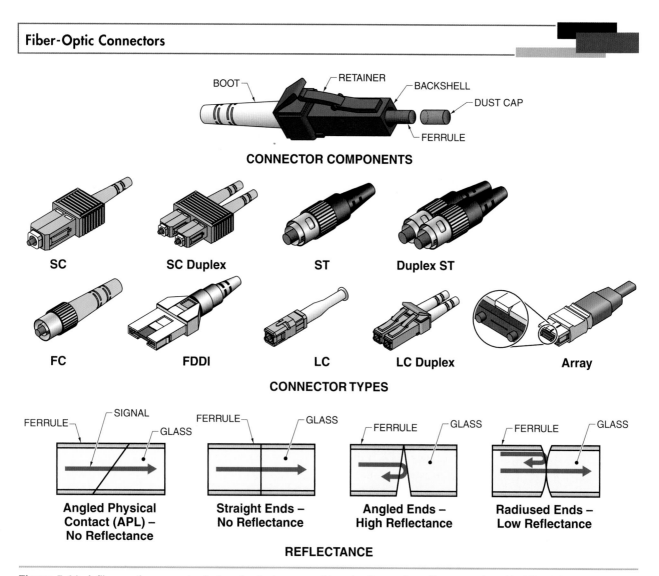

Figure 5-11. A fiber-optic connector is the physical means of terminating a glass fiber strand and holding it in perfect alignment with another glass fiber strand or the receiving hardware.

Adhesive-Type Connectors. Fiber can be glued into position within a connector with a heat-cured, UV-cured, or anaerobic-cured adhesive. With heat-cured adhesive connectors, the adhesive is injected into the connector assembly, and the connector then placed in an oven to cure and harden the adhesive. Because an oven is required to terminate these connectors, the process can be costly and time consuming. Typical cure times are about 20 minutes.

With UV-cured adhesive connectors, the adhesive is injected into the connector, and a UV light source is used to cure it. Although this requires the use of a UV light source, the process can be completed in about a minute, making it more effective than the heat-cured process.

With anaerobic-cured adhesive connectors, a two-part epoxy adhesive-and-hardener system is used to cure the adhesive. With this system, a resin is inserted into the ferrule and a hardener applied to the fiber. When the fiber is inserted into the resin-filled ferrule, the hardener reacts with the resin and cures the adhesive in about 15 seconds. No additional equipment is required.

Field-Polish Connectors. A *field-polish connector,* also known as a hand-polish connector, is a fiber-optic connector that requires hand-polishing of the glass protruding from the end of the connector ferrule. Typically, field-polish fiber-optic connectors are glued to a strand of glass with an anaerobic two-part epoxy adhesive developed for the purpose of terminating fibers. After stripping the fiber buffer and coating and cleaning it, it is inserted into the end of the connector until it protrudes from the connector ferrule. A fiber-optic scribe is then used to score the glass at the end of the ferrule. A fiber scribe is a low-cost type of fiber cleaving tool. A cleaved fiber leaves a clean, 90° cut.

After scribing the glass, the fiber is gently pulled to remove the excess protruding from the ferrule. The connector is then inserted into a polishing puck and hand-polished against a fiber-polishing abrasive sheet in a figure-eight motion to make it flat

and free of irregularities. Once complete, the light pulse can travel through it without loss or distortion. **See Figure 5-12.**

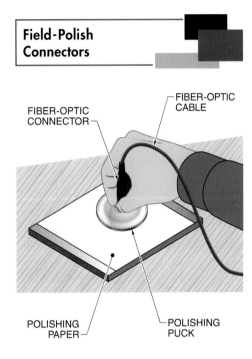

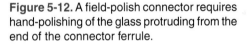

Figure 5-12. A field-polish connector requires hand-polishing of the glass protruding from the end of the connector ferrule.

The process of gluing, scribing, and hand-polishing each connector is a tedious, time-consuming, and difficult process. A technician usually requires many hours of practice to become proficient and consistently produce high-quality results. Also, due to the time it takes to become proficient with hand-polishing connectors, many are discarded on job sites as they are improperly scribed or polished, which increases overall material costs. While field-polish connectors are becoming less common, they are still used in applications to match existing installations. Or sometimes, specifications require a lower loss connection without the internal splice of a mechanical connector, such as those found in prepolished connectors.

Splice-Type Connectors. A *splice* is the joining of two or more fiber-optic strands. A *crimp-on connector* is a fiber-optic connector that is crimped together with a crimping tool to complete the mechanical splice inside the connector. When joining two cables together, splices that involve scribing and index-matching material are commonly used. An *index-matching material* is a material in liquid, paste, film, or gel form that has a refractive index that is nearly equal to the core index of a fiber and is used to reduce reflections from a fiber end face. The most common index-matching material used with fiber-optic connectors is in gel form.

Another type of splice is the fusion splice. The fusion splice technique uses a tool to melt the ends of two glass strands together and make them a single length of glass. While this type of splice results in lower signal loss, it is more costly to perform than splicing fibers with an index-matching gel.

The newest type of connector is known as a fuse connector. It uses fusion-splice technology with a factory-polished connector (commonly referred to as a pigtail). Fuse connectors have the advantages of extremely low signal loss due to the integrity of the fusion splice and factory-polished end. It also has the advantage of being relatively easy to install. The disadvantage may be the cost, although this will likely improve in time. Installation of these types of connectors also requires the use of a fusion splicer, which is a fairly expensive tool.

Prepolish Splice-Type Connectors. Prepolish splice-type connectors have prepolished fibers and have become the industry standard due to the ease of termination and consistent signal readings. A prepolish splice-type fiber-optic connector has a fiber stub inserted in the ferrule that is mechanically polished at the OEM's factory rather than manually in the field by a technician. Therefore, it does not require the slow process of field-polishing each connector with a polishing puck. When terminating prepolish splice-type connectors, the field fiber is cut by a technician using a fiber scribe. The field fiber is then inserted into the connector and seated in alignment with a fiber stub that extends to the polished surface of the ferrule. A small drop of index-matching gel between the field fiber end and the fiber stub end helps make the transition from one fiber to the other with a minimum amount of signal loss. The connector body and cover is then snap-fit together to complete the termination process. **See Figure 5-13.**

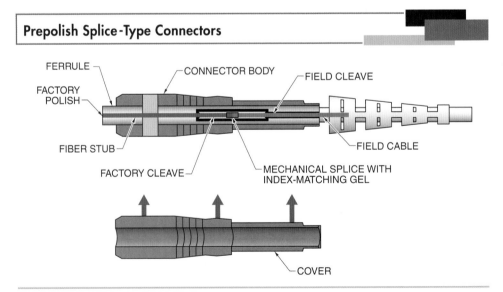

Prepolish Splice-Type Connectors

FERRULE

FACTORY POLISH

CONNECTOR BODY

FIELD CLEAVE

FIBER STUB

FACTORY CLEAVE

MECHANICAL SPLICE WITH INDEX-MATCHING GEL

FIELD CABLE

COVER

Figure 5-13. A prepolish splice-type fiber-optic connector has a fiber stub inserted in the ferrule that is mechanically polished at the OEM's factory rather than manually in the field by a technician.

FIBER-OPTIC SAFETY

A fiber-optic cable consists of strands of coated glass that are covered by a protective sheath designed for the environment where it is installed. While protected within the cable sheath, the cable is both safe from damage and safe to handle. When terminating cables and connectors however, the outer layers of the cable sheath and strength members are stripped away so that the glass can be terminated in a connector or spliced to another fiber. Because a standard fiber-optic strand is only about the size of a human hair, it can break easily. A fiber strand is also difficult to see and can be sharper than a needle.

Extreme care is essential to safety when handling fiber cables, patch cords, and connectors. Eye protection is always required in the job of a VDV technician, but it is especially so when handling fiber. Since fiber strands are difficult to see, small pieces of glass can easily get in the eyes if proper safety procedures are not practiced. In addition to wearing protective eyewear, a technician must avoid touching the eyes or face while working with or near fiber-optic components and devices. Basic fiber-optic safety practices include the following:

- Keep all food and beverages out of the work area. If fiber particles are ingested, they can cause internal hemorrhaging.
- If possible, perform work on black work surfaces to help visually locate fiber scraps.
- Wear disposable aprons to reduce the amount of fiber particles on clothing. Fiber particles on clothing can later get into food, drinks, or be ingested by other means.
- Always wear safety glasses with side shields and protective gloves. Treat fiber-optic splinters as glass splinters.
- Never look directly into the end of fiber cables until verifying that there is no light source at the opposite end. Use a fiber-optic power meter to make certain the fiber is dark.
- Only work in well-ventilated areas.

- If wearing contact lenses, do not handle lenses until thoroughly washing hands with soap and water.
- Never touch eyes or skin near eyes while working with fiber-optic systems until work clothing has been removed and hands have been thoroughly washed with soap and water.
- Keep all combustible materials away from curing ovens.
- Put all cut fiber pieces in a properly marked container for disposal.
- Thoroughly clean the work area when finished.
- Do not smoke while working on fiber-optic systems.

Tech Tip

A single-mode connector can be used with a multimode fiber, but a single-mode fiber cannot be used with a multimode connector. Single-mode connectors are manufactured to tighter tolerances than multimode connectors, making it impossible to use them with multimode fibers.

FIBER-OPTIC DIGITAL TRANSMISSIONS

By using ASCII code in twisted-pair copper systems, messages are constructed with voltage levels by using 1s and 0s. The same code is also used in fiber-optic systems. However, rather than using voltage to represent a 1 or 0, fiber-optic systems use light pulses. A light pulse "on" represents 1, and a light pulse "off" represents 0. Because a fiber-optic system does not use voltage and there is no metal in the conductor, a fiber-optic cable has distinct advantages and disadvantages when compared to twisted-pair copper cable. In other words, copper and fiber-optic cables serve the same purpose but are different physically. **See Figure 5-14.**

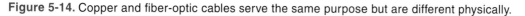

Copper vs Fiber-Optic Cables	
Copper	Fiber Optics
• Low material cost	• No metal materials required
• Low electronic-component cost	• No noise interference
• Ease of installation	• No lightning protection requirement
• Less use of specialized tools and test equipment	• More secure (cables cannot be intrusively tapped or spliced)
• Sturdy construction	• More lightweight
• More commonly used	• Longer communication distances possible with a single run
	• Higher transmission speeds

Figure 5-14. Copper and fiber-optic cables serve the same purpose but are different physically.

When transmitting light, the same pattern of light pulses that enter the transmission end should exit the receiving end. The receiving end can be in a different room, in a different building, or miles away. The arrangement of a transmitter on one end and a receiver on the opposite end makes fiber-optic data transmission essentially a "one-way street." In order for two-way communication to be possible, there must be an additional fiber-optic strand working in the opposite direction. Fiber-optic systems almost always have a pairing system of a transmitter and receiver to accommodate two different fibers. **See Figure 5-15.**

Note: One exception to the pairing system design is a closed-circuit camera system, which only transmits and does not receive.

Optical Transmitters and Receivers

An *optical transmitter* is a device that takes an electrical input and converts it to an optical output. The most common type of optical transmitter is a light-emitting diode (LED). An LED is a semiconductor device that emits light when an electric current is passed through it. The intensity of the light varies with the amount of current passing through it. An LED is the least expensive type of optical transmitter and is usually used with multimode fiber.

A *laser transmitter* is a device that takes an electrical input and converts it to a concentrated light output. Light output from lasers are concentrated in a small point, while LED light output is transmitted over a larger area. There are two common types of laser transmitters. The fabry-perot (FP) laser was developed first. It has excellent performance but is extremely costly to produce. More recently, OEMs have developed a vertical cavity surface-emitting laser (VCSEL) which is more cost effective. In practice, the FP laser is usually just referred to as a laser, and the VCSEL is pronounced "vix-ill."

All three transmitters serve the same function. They convert electrical signals into pulses of light. An LED provides a wider, more diverging light pattern than a laser light source. A wide light pattern makes it more difficult to couple light into a fiber-optic cable and limits the bandwidth. Lasers provide a narrower, less diverging light pattern that allows faster speeds and the use of single-mode fiber. **See Figure 5-16.**

An *optical receiver* is a device that converts optical signals back into a replica of the original electrical signal. It serves the opposite function of an optical transmitter. Optical receivers are also known as optical sensors. One type of optical receiver is the photodiode. A *photodiode* is a semiconductor device that detects light and converts it into an electrical signal.

A *transceiver* is a device that both transmits and receives. Many modern fiber-optic applications are being developed to use transceivers.

Fiber-Optic Data Transmission

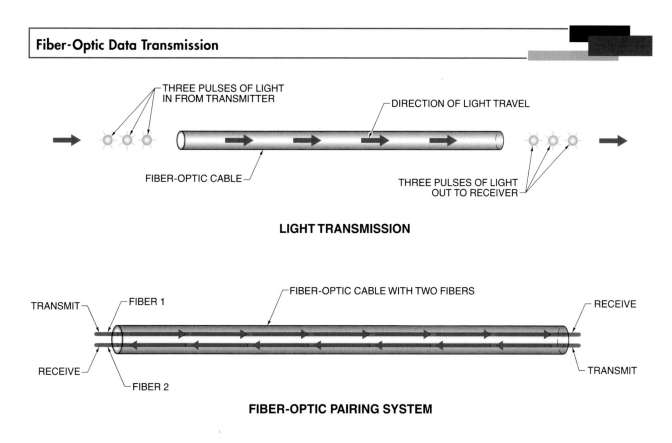

LIGHT TRANSMISSION

FIBER-OPTIC PAIRING SYSTEM

Figure 5-15. Fiber-optic systems almost always have a pairing system of a transmitter and receiver to accommodate two different fibers.

LED vs Laser Light Source

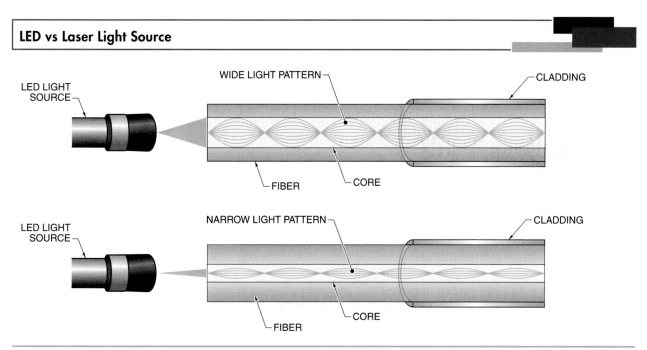

Figure 5-16. An LED provides a wider, diverging light pattern, while a laser provides a narrower, less diverging light pattern that allows faster speeds and the use of single-mode fiber.

Cable Bend Radius

Fiber-optic cables function as a guide for light. A light source is used to transmit a pulse of light at one end of the cable. The light then travels along the length of the cable until it reaches the opposite or receiving end. The cable itself can take a varied path, including bends. Because of this, fiber-optic cables have a bend radius requirement. A bend radius is the minimum radius that a cable can bend without causing damage. A bend radius is usually expressed as a multiple of the diameter of the cable and is measured from the inside as the radius arc. Failure to maintain the proper bend radius can cause signal loss or glass breakage within the cable. With fiber-optic cable, the standard for minimum bend radius is no less than ten times the cable diameter.

A fiber-optic cable can generally endure an extremely high amount of pulling, or tensile, force (up to 700 lb for some types of outdoor cable), but it is sensitive to being bent past its bend radius. A sharp bend in a fiber-optic cable results in distortion of the light signal passing through the glass or damage to the glass strand, such as breakage. Cables that have broken glass strands are extremely expensive to repair. Usually, a new section of cable must be spliced to the unbroken glass on each side of the broken strand. This process is time consuming and increases the amount of optical loss in the cable. For this reason, cables with broken glass strands are often replaced rather than repaired.

Bends can be macrobends or microbends. A *macrobend* is a bend that occurs when a fiber-optic cable is bent around a large radius. A *microbend* is a small distortion in a fiber-optic cable caused by a crushing or pinching force. Microbends change the path of the signal, causing it to be absorbed by the cladding. Sometimes, a microbend is not visible to the naked eye.

The proper bend radius is measured from the inside of the bending arc, rather than the outside. **See Figure 5-17.** A multimode fiber is less sensitive to an incorrect bend radius than a single-mode fiber. However, some cable OEMs have developed bend insensitive cable that allows a cable to be bent in the "kink zone" and keep its integrity intact. The "kink zone" is the area of the cable that is most susceptible to an incorrect bend radius.

LIGHT INTENSITY, PROPAGATION, AND MEASUREMENT

A fiber-optic path between a transmitter and receiver is the physical connection between them. The measure of the performance of that path involves the intensity, or brightness, of the light entering and exiting the fiber. The propagation of light refers to the process of light travel from one end of the fiber to the other, including any physical parameters that would affect the light intensity. The measure of how well a fiber-optic cable works is the amount of signal loss, or attenuation, that occurs as the light travels through the fiber. A small amount of light is lost along the path of every cable. Fiber-optic cable standards have a standard signal loss per length for their products. Typically, this is stated in decibels of loss per meter of cable. Signal strength is also lost at every splice and connector termination point.

A *loss budget* is the calculated signal loss that is expected for a fiber-optic installation. A typical project loss budget is calculated by first multiplying the cable length by the rate of signal loss per unit of length. The product is the amount of cable loss. Then the number of splices and connector pairs are likewise multiplied by the amount of signal loss per device. All three products are then added, and the result is the total loss budget. This loss budget must be less than the operating budget of the transmitter-receiver pairing. The loss and operating budgets are also defined by the project's standards. **See Figure 5-18.**

FIBER-OPTIC CABLE WARRANTIES

Fiber-optic cable manufacturers need certified installers to install and test their products. Although cables are tested prior to shipping, cables are tested again after installation as proof of proper system operation in order to offer long-term operational warranties. All OEMs have registration and warranty programs to help sell, service, and properly install their products. A contractor usually becomes certified by a specific OEM by completing a training course and signing a contract to agree to provide service for a warranty period of 20-to-25 years. A warranty does not include any mishandling or damage after installation. Once installed and tested, a fiber-optic cable rarely fails unless it is disturbed. For this reason, warranty claims on fiber-optic cables are extremely rare. In order to entice a contractor to be loyal to an OEM, there is usually a rebate program based on the total amount of fiber-optic product purchased in a year.

Fiber-Optic Cable Bend Radius

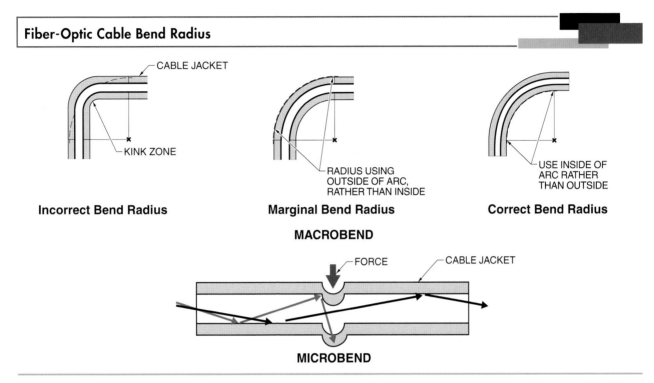

Figure 5-17. A proper bend radius must be maintained at all times with a fiber-optic cable. A sharp bend can result in distortion of the light signal passing through the glass or damage to the strand.

Loss Budget Calculations

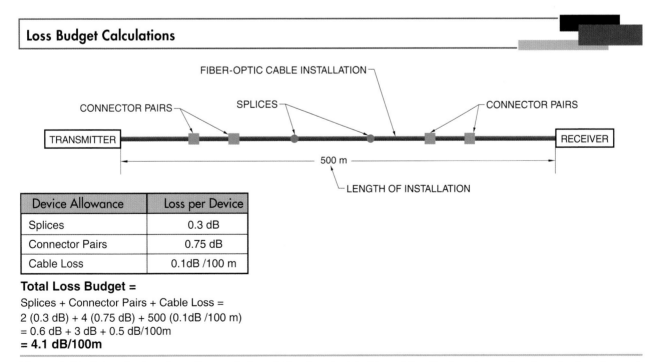

Device Allowance	Loss per Device
Splices	0.3 dB
Connector Pairs	0.75 dB
Cable Loss	0.1dB /100 m

Total Loss Budget =
Splices + Connector Pairs + Cable Loss =
2 (0.3 dB) + 4 (0.75 dB) + 500 (0.1dB /100 m)
= 0.6 dB + 3 dB + 0.5 dB/100m
= 4.1 dB/100m

Figure 5-18. A loss budget for a fiber-optic installation is calculated by adding the loss for each splice, the loss for each connector pair, and the cable loss.

Summary

Fiber-optic cabling provides an alternative to copper-structured cabling or complements an overall system that includes copper. It works because it transmits information, not power. Fiber optics involve lightweight cables that are unaffected by electromagnetic interference. Because of the advantages, fiber-optic cabling is quickly becoming the standard for most high-bandwidth communication applications. Fiber types used in VDV systems are primarily single-mode but can be multimode as well. Key performance factors include attenuation, acceptance angle, numerical aperture, and modal and chromatic dispersion. Several different types of cables and connectors can be used to install fiber-optic systems. Fiber-optic safety requires specific care when handling glass fibers. The main difference between fiber-optic cables and copper cables is that copper transmits data through electrical signals but fiber optics transmit data through the use of light pulses.

Chapter Review

1. Explain the main difference between the operation of a fiber-optic communications system and a copper based communications system.

2. List six advantages of fiber-optic systems as compared to copper systems.

3. What are the three parts of a glass fiber strand?

4. What is the typical cladding diameter of a fiber-optic strand used in a VDV system?

5. Why do OEMs use coatings of different colors on individual fibers?

6. Explain the difference between a cable jacket and a cable sheath.

7. What is a ripcord?

8. Explain the difference between a multimode fiber and a single-mode fiber.

9. What is the core diameter for most modern multimode installations?

Chapter Review

10. What type of light sources are used with multimode and single-mode cables?

11. List six factors that cause attenuation.

12. What is bandwidth?

13. How does the NEC® designate cables as conductive or nonconductive?

14. Explain the difference between loose-tube and tight-buffered cable.

15. What is the color of a 50 µm multimode cable jacket or sheath?

16. Explain what could happen if a glass fiber is not perfectly scribed and aligned.

17. List and describe the five elements of a fiber-optic connector.

18. List the four main criteria used to measure fiber-optic connector performance.

19. List five common fiber-optic connector styles.

20. Explain where a field-polish connector may be used.

21. List 12 basic fiber-optic equipment safety practices.

22. What happens when a ray of light meets the boundary between the core and the cladding in a fiber-optic strand and why?

23. What is a bend radius?

Chapter Review

24. What is a macrobend?

25. How is a typical loss budget calculated?

Chapter Activity   Fiber Optic-Cable Systems

Provide the NEC® acronym for the following types of fiber-optic cable:

1. Low-smoke density

2. Optical fiber conductive

3. Riser-rated for vertical runs

4. Optical fiber nonconductive

5. General purpose

6. Plenum-rated for use in air handling plenums

Chapter 6

VDV Prints

As cable and satellite TV, wiring and cabling, the Internet, and personal computers have become part of daily life, the need for low-voltage wiring to connect these devices has grown rapidly. VDV systems provide connections for telephones (voice); computers (data); and cable and satellite TV (video).

The size, complexity, and type of construction project (residential or commercial) will determine the number and type of prints used to depict a system. For residential installations, a basic system is usually shown on a residential floor plan along with the power and lighting systems. For commercial installations, VDV systems are typically shown individually on a separate, dedicated set of prints. Commercial VDV system prints are typically divided into riser diagrams, floor plans, and detail drawings. In addition to VDV-specific prints, other electrical prints are referenced as required.

OBJECTIVES

- Describe the meanings of abbreviations, symbols, and legends associated with specific prints.
- Explain the differences between riser diagrams and floor plans.
- Explain the differences between commercial and residential floor plans.
- Describe detail drawings for equipment rooms, VDV outlets, firestopping, and conduit fill.

Digital Resources

ATPeResources.com/QuickLinks
Access Code 838502

VDV SYSTEM PRINTS

The information on VDV systems tends to be grouped together because these systems are typically installed by the same contractor. Other names for VDV systems include structured cabling, telecommunications wiring, and IT infrastructure.

The size, complexity, and type of construction project (residential or commercial) determine the number and type of prints needed to depict a VDV system. A basic residential VDV system is typically shown on a residential floor plan along with power and lighting information. A large commercial VDV system is typically shown on a separate set of prints. These VDV system prints are typically composed of riser diagrams, floor plans, and detail drawings. **See Figure 6-1.**

For complete information on a VDV system, other electrical prints may need to be consulted as well. For example, an electrical site plan may contain information on the conduit and cables connecting a building to the local telephone company or to a customer-owned outside plant. However, it is important that VDV systems be installed separately from most other power systems involved with a project because AC frequencies will cause interference with VDV signals.

VDV Abbreviations, Symbols, and Legends

VDV prints include lists of abbreviations and symbols. The abbreviations and symbols are used throughout the prints and provide a clear and concise method to convey information without cluttering the diagrams. **See Figure 6-2.** The abbreviations are specific to cable television, telecommunications, and structured cabling wiring systems. **See Appendix.**

Typically the symbols used on VDV prints are part of a standard, such as one of those regulated by the American National Standards Institute (ANSI). The symbols depict outlets for data, telephone, television, and VDV-related equipment. The symbols are often modified with letters to provide additional information. For example, the letter "F" may be used to indicate a flush-mounted outlet, and the letter "S" may indicate a surface-mounted outlet.

Some VDV designers prefer to use a custom set of symbols. While an OEM-developed set of symbols may resemble a set of standard VDV symbols, OEM-developed symbols often have minor differences. For example, outlet address information may be included with OEM-developed symbols. An *outlet identifier* is a unique system of letters and/or numbers used for identifying cables and their locations in a VDV system. The outlet address format used is generally specified by the VDV designer, customer, or standards.

A *legend* is a description or explanation of symbols and other information on a drawing or print. Most VDV prints include legends, which also provide information about the installations. A legend may be included with VDV abbreviations and symbols, or it may be on a separate print. Information provided in a legend may apply to multiple installations throughout the VDV prints. For example, the size and location of pipe sleeves would be included in a legend.

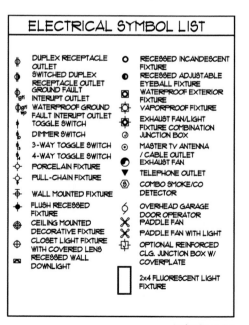

Linden Group, Inc.

All prints include an electrical symbol list, specific to each project or installation.

VDV System Prints

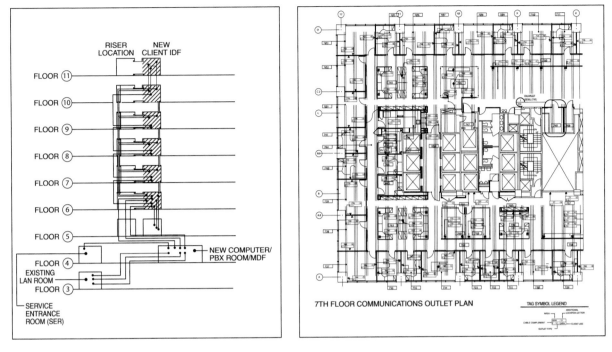

VDV RISER DIAGRAM

VDV FLOOR PLAN

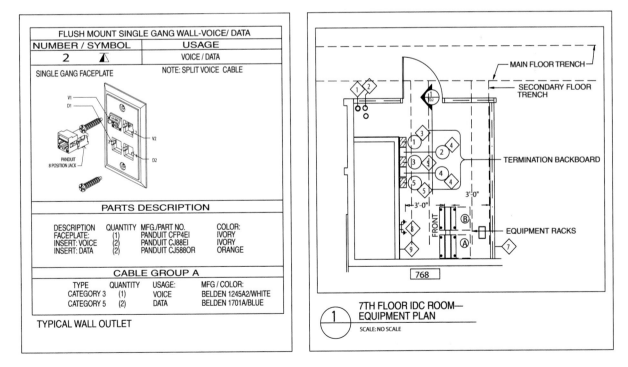

VDV DETAIL DRAWINGS

Figure 6-1. VDV system prints are typically composed of riser diagrams, floor plans, and detail drawings.

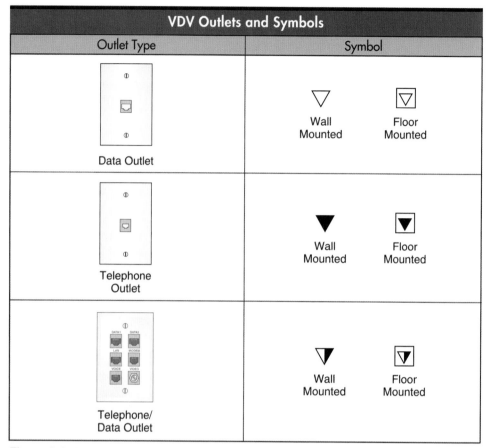

Figure 6-2. Specific symbols are used to represent telephone outlets and data outlets in VDV systems.

Riser diagrams are provided for VDV installations in multistory building construction projects and are available as part of the project's prints.

Each item in a legend is associated with a number or letter. Often the number or letter is shown inside a geometric shape, such as a circle, diamond, or octagon. The number or letter may appear throughout the various VDV prints where that specific piece of information is relevant. **See Appendix.** VDV print legends also include specification notes, VDV equipment model identification, and installation locations. **See Figure 6-3.**

VDV Riser Diagrams

Riser diagrams are normally provided for VDV installations in multistory buildings. Multistory buildings can be commercial office spaces, high-rise residential condominiums, retail stores, or mixed-use buildings (a combination of office, residential, and retail spaces). Riser diagrams are not required for multistory, single-family residences.

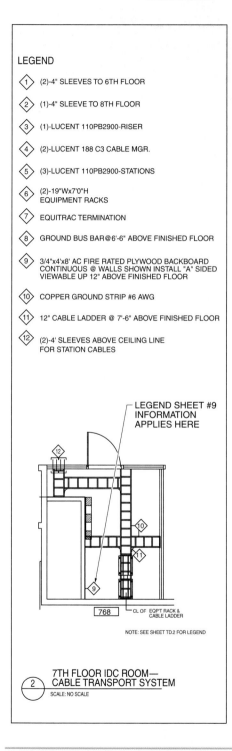

VDV Legend Information

LEGEND

1 (2)-4" SLEEVES TO 6TH FLOOR

2 (1)-4" SLEEVE TO 8TH FLOOR

3 (1)-LUCENT 110PB2900-RISER

4 (2)-LUCENT 188 C3 CABLE MGR.

5 (3)-LUCENT 110PB2900-STATIONS

6 (2)-19"Wx7'0"H
   EQUIPMENT RACKS

7 EQUITRAC TERMINATION

8 GROUND BUS BAR @ 6'-6" ABOVE FINISHED FLOOR

9 3/4"x4'x8' AC FIRE RATED PLYWOOD BACKBOARD
   CONTINUOUS @ WALLS SHOWN INSTALL "A" SIDED
   VIEWABLE UP 12" ABOVE FINISHED FLOOR

10 COPPER GROUND STRIP #6 AWG

11 12" CABLE LADDER @ 7'-6" ABOVE FINISHED FLOOR

12 (2)-4' SLEEVES ABOVE CEILING LINE
    FOR STATION CABLES

LEGEND SHEET #9
INFORMATION
APPLIES HERE

768   CL OF EQP'T RACK &
       CABLE LADDER

NOTE: SEE SHEET TD.2 FOR LEGEND

2 7TH FLOOR IDC ROOM—
   CABLE TRANSPORT SYSTEM
   SCALE: NO SCALE

Figure 6-3. Each item on a legend is associated with a number or letter. The number or letter is used on the prints to avoid cluttering the diagrams.

A VDV installation in a multistory building has backbone cabling and horizontal cabling. Backbone cabling is routed from a main distribution frame (MDF) in a telecommunications room to an intermediate distribution frame (IDF). An *intermediate distribution frame (IDF)* is a metal rack designed to connect cables and is located in an equipment room or telecommunications room. Changes in wiring to a system are typically performed at the IDF. Horizontal cabling is routed from the information outlets at individual work stations to an IDF. Both the backbone and horizontal cables are cross connected in the IDF and MDF.

VDV riser diagrams provide an easy-to-understand overview of the backbone cables in a commercial building. **See Figure 6-4.** A VDV riser diagram depicts the cables that enter the building from the local telecommunications provider into the service entrance room and the cables that are routed from the MDF to the IDF on each floor of the building. For a new construction project, all floors in the building are depicted in a VDV riser diagram. However, for a retrofit project on an existing building, a VDV riser diagram may omit floors that are not involved in the retrofit.

In addition to VDV cables, the installation may require a grounding system. Telecommunication equipment and cables that require grounding are connected to a dedicated telecommunications bonding backbone. The bonding backbone may also be depicted in a VDV riser diagram.

A VDV riser diagram is not drawn to scale but does designate MDFs, IDFs, the types of cables run (copper or fiber optic), and the size of cables run (number of pairs for copper or number of fibers for fiber optic). The exact route of the backbone cable is not shown, and VDV riser diagrams usually do not give the size of the conduit or raceway for the backbone cables. Electrical prints must be consulted for information on conduit or raceway size.

VDV Riser Diagrams

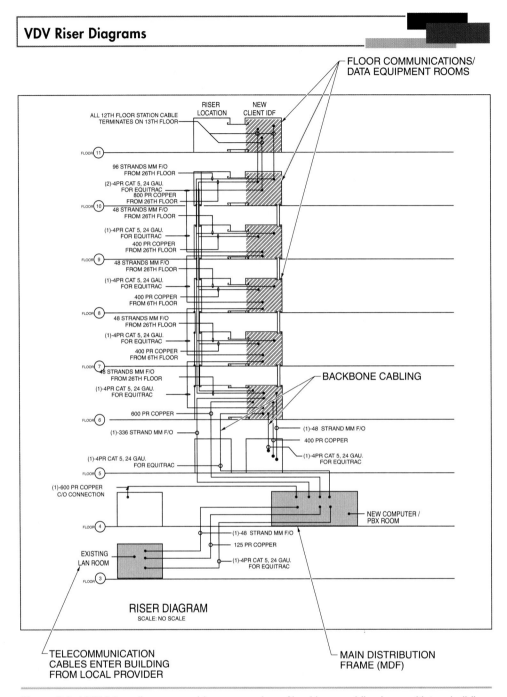

Figure 6-4. A VDV riser diagram provides an overview of backbone cabling in a multistory building.

VDV Floor Plans

A *plan view* is a drawing of an object as it appears looking down from a horizontal plane. A *floor plan* is a drawing that gives a plan view of each floor of a building. A floor plan shows cellular metal floor raceways, elevator shafts, equipment locations, doors, plumbing fixtures, rooms, stairways, walls, windows, and modular furniture partitions as well as power poles to feed the partitions. Typically a description and/or room number is shown for each room, including closets and corridors.

A VDV floor plan shows the location of VDV outlets and modular furniture partitions. **See Figure 6-5.** VDV floor plans are drawn for both commercial and residential projects. Although VDV floor plans for commercial and residential projects contain much of the same information, there are significant differences between them.

Commercial VDV Floor Plans. Commercial VDV floor plans are drawn for buildings such as commercial office buildings, financial institutions, schools, and retail stores. Large commercial projects have a separate floor plan for VDV systems and are drawn for each floor of the project. More than one VDV floor plan may be necessary if a floor is extremely large. Small commercial projects normally include the VDV elements on the power floor plans or architectural plans, similar to residential projects.

VDV Floor Plans

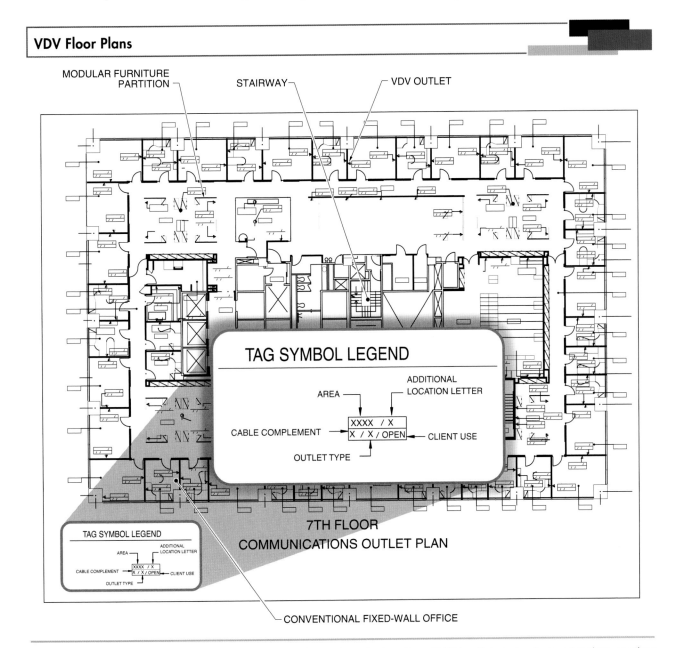

Figure 6-5. VDV floor plans for large commercial projects show the location of VDV outlets but not power outlets or other devices typically shown on floor plans.

Typically, most VDV outlets in a commercial building are mounted at the same height and their location determined from the print. The prints or specifications provide the standard mounting height and whether the outlets are to be installed exactly per the scaled dimension or to the nearest stud. Normally, dimensions are only shown for outlets that are not typical, but some prints provide exact dimensions for every outlet.

In addition to the location of VDV outlets, commercial floor plans may contain sheet notes. **See Figure 6-6.** Sheet notes provide information about outlet addresses and installation of components.

Tech Tip

In rooms where wireless signals are weak, a cable may be routed from a router to a new VDV outlet.

Commercial VDV Floor Plans

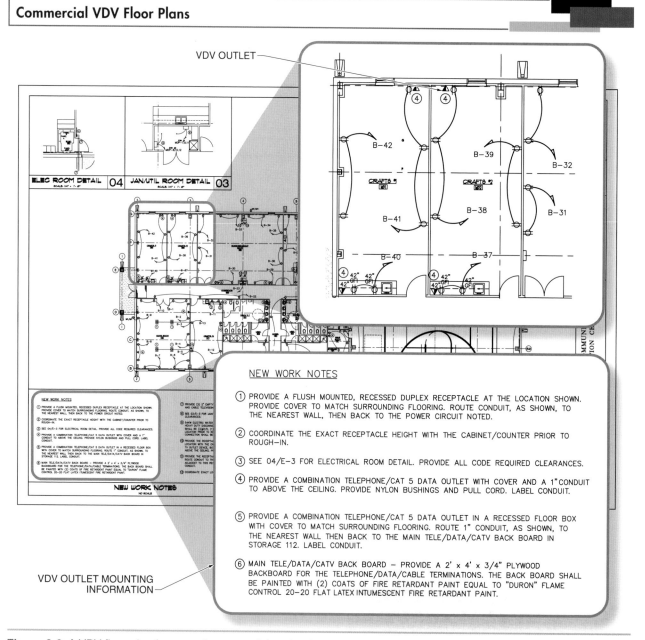

Figure 6-6. A VDV floor plan for a small commercial project is typically combined with the building's power floor plan.

Unlike power devices and outlets on a power floor plan, VDV outlets on a VDV floor plan do not have lines representing communication cables drawn from each outlet to an IDF or MDF. The number and type of cables from each outlet can be found in the specifications, as part of the outlet symbol or as a sheet note.

For example, commercial VDV installation could consist of up to four unshielded twisted-pair (UTP) cables routed from an IDF to each outlet. Each outlet has four separate jacks, and the UTP cables are used for voice and/or data. **See Figure 6-7.**

Residential VDV Floor Plans. Usually, there is only one project floor plan for a residential construction project. The residential VDV information is shown on the project floor plan. As with a commercial VDV floor plan, a residential project floor plan provides the location of VDV outlets. **See Figure 6-8.** Unlike a commercial power floor plan however, a residential floor plan also includes the following information:

- electrical outlet and device locations
- light fixture and switch locations
- interior and exterior building and room dimensions
- HVAC register locations
- plumbing fixture information
- door and window information
- floor and wall covering information
- floor and/or roof-framing information

As with the VDV outlets in a commercial project, most VDV outlets in a residential project are mounted at the same height, and their location is determined from the print. Information on VDV outlet mounting height, and whether mounting locations are exactly per the scaled dimension or to the nearest stud, is detailed in the print notes or specifications. Unlike a commercial VDV print, VDV outlet address information is not provided.

Tech Tip

In most new home construction, VDV cables have a dedicated access panel, similar to cables in electrical systems.

VDV UTP Cables

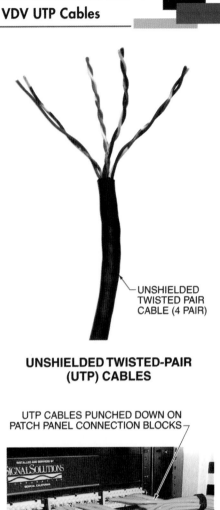

UNSHIELDED TWISTED PAIR CABLE (4 PAIR)

UNSHIELDED TWISTED-PAIR (UTP) CABLES

UTP CABLES PUNCHED DOWN ON PATCH PANEL CONNECTION BLOCKS

PATCH PANEL UTP CABLES

Figure 6-7. Unshielded twisted-pair (UTP) cables used for voice and data transmission terminate at connector blocks on the rears of patch panels.

Residential VDV Floor Plans

EXTERIOR DIMENSION INFORMATION

VDV DATA OUTLET

HVAC REGISTER

ELECTRICAL OUTLET

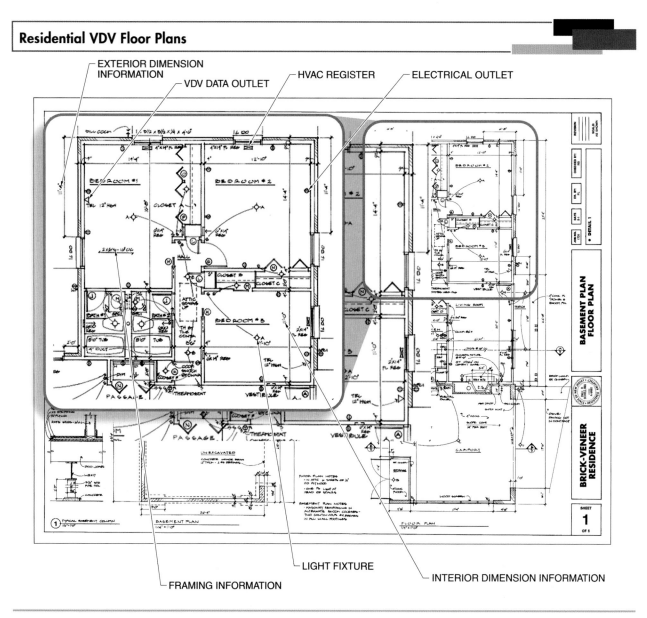

LIGHT FIXTURE

INTERIOR DIMENSION INFORMATION

FRAMING INFORMATION

Figure 6-8. A residential VDV floor plan is part of a project floor plan and includes power, lighting, and other electrical, mechanical, and architectural elements.

Some residential VDV installations can consist of two UTP cables and two coaxial cables routed from a distribution cabinet (network center) to a single outlet in each room. VDV outlets are available to accommodate multiple jacks of various configurations for voice, data, or video. The distribution cabinet serves as the connection point for cables from the local telephone service provider and local cable provider, and the UTP and coaxial cables routed to various rooms in the residence.

The UTP cables are used for voice or data, and the coaxial cables are typically used for video. **See Figure 6-9.** Some residential projects also include cable for speakers and home theater systems.

Certain cable OEMs combine the two UTP and two coaxial cables under a common insulation jacket to create a single hybrid cable for residential applications. **See Figure 6-10.** This hybrid cable decreases installation time and helps with cable identification.

Residential VDV Installations

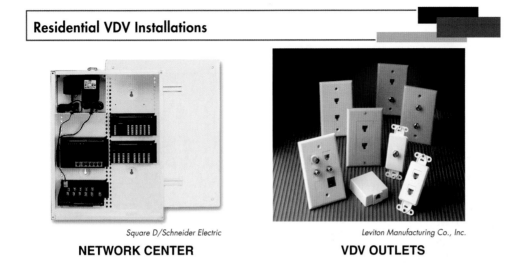

Square D/Schneider Electric

NETWORK CENTER

Leviton Manufacturing Co., Inc.

VDV OUTLETS

Figure 6-9. A common residential VDV installation consists of UTP and coaxial cables that run from an IDF to each room with an outlet. Each outlet may have multiple jacks of various configurations for voice, data, and/or video.

Hybrid Cables

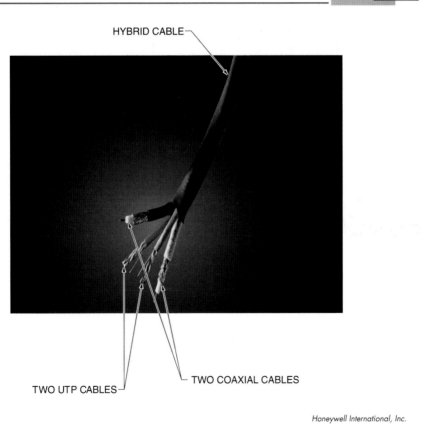

HYBRID CABLE

TWO UTP CABLES

TWO COAXIAL CABLES

Honeywell International, Inc.

Figure 6-10. A hybrid cable has two individual UTP cables and two individual coaxial cables under a common insulation jacket and is used for residential applications for ease of installation.

VDV Detail Drawings

VDV detail drawings provide additional information that is not shown on VDV riser diagrams or VDV floor plans. The number and type of detail drawings depends on the size of the construction project. A small commercial office building may have a few detail drawings, while a large campus with multiple buildings would have several detail drawings. VDV detail drawings are rarely found on prints for single-family residences. The most common types of detail drawings include equipment room detail drawings, VDV outlet detail drawings, firestopping detail drawings, and conduit fill detail drawings. The VDV detail drawings provide a high degree of specificity to ensure that the VDV installation meets applicable codes and the requirements of the customer.

Equipment Room Detail Drawings. A detail drawing of a main distribution frame shows backbone and horizontal cabling, connector blocks, telephone equipment, network equipment, and cable management and support devices. Several equipment room detail drawings may be required for complex installations. VDV technicians must consult equipment room detail drawings before installing any equipment.

Equipment room detail drawings are required to show the location and mounting method for VDV equipment. The drawings indicate where and how raceways are to be mounted in ceiling spaces and between floors. Elevation and plan views are used to identify the location of cable ladders, equipment racks, termination backboards, connector blocks, and raceways. **See Figure 6-11.** A legend is included to avoid clutter on the drawing.

Tech Tip

Equipment rooms must have environmental controls to provide HVAC and air filtration at all times. A temperature range of 64°F to 75°F (18°C to 24°C) and 30% to 55% relative humidity must be constantly maintained.

Equipment Room Detail Drawings

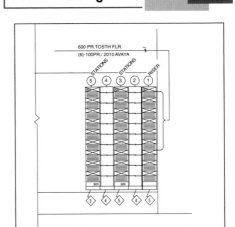

ELEVATION VIEW

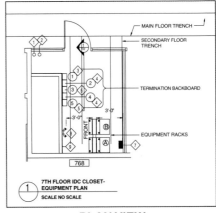

PLAN VIEW

LEGEND

Figure 6-11. Equipment room detail drawings include elevation views, plan views, and legends that show the locations of cable ladders, equipment racks, termination backboards, connector blocks, and raceways.

VDV Outlet Detail Drawings. There are many different varieties of VDV outlets on any construction project. The variations include the following:

- Mounting style—wall mount, floor mount, or modular furniture mount
- Single or double gang outlets
- Number of jacks per outlet and types, either telephone or data (with two telephone jacks and two data jacks commonly found in commercial installations)

VDV outlet detail drawings provide comprehensive information for each type of outlet. Information includes the exact termination scheme for a UTP cable to the jacks, the outlet addressing format, the outlet symbol, and the number and type of cables and jacks. Some VDV outlet detail drawings include the part numbers for cables, faceplates, and jacks. **See Figure 6-12.** Additional information found in VDV outlet detail drawings includes precise information on the type of label used for outlet faceplates, such as the size of the label, the label font, and the label font size.

Tech Tip

In some larger offices, it is helpful to install two outlets on opposite walls at each location. This will help to keep patch cord lengths and cable clutter to a minimum.

VDV Outlet Detail Drawings

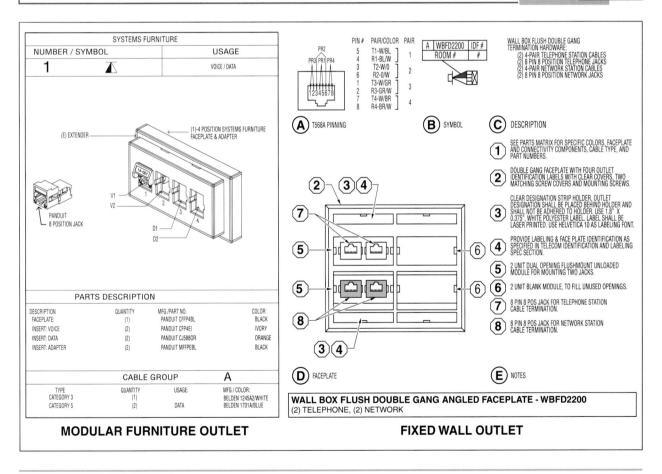

Figure 6-12. VDV outlet detail drawings provide information on the number and type of jacks, the termination scheme for the UTP cable to the jacks, and labeling for each type of VDV outlet.

Firestopping Detail Drawings. A *firestop* is a system made of various fire-rated components used to inhibit the spread of fire. *Firestopping* is the process of applying and installing a material or member that seals an opening in a fire-rated wall, floor, or membrane to inhibit the spread of fire, smoke, and fumes in a structure. Holes and penetrations in fire-rated walls and floors must be sealed to maintain the fire rating of the initial installation. Firestopping is an extremely important element of any VDV project and is typically performed with firestopping caulk. *Firestopping caulk* is a material commonly used to seal holes drilled through walls or structural members to prevent the spread of fire. It is available as expandable or heat-absorbing.

There are many variables to consider when choosing a firestopping system. The variables include the initial rating of the wall or floor; the material the wall or floor is constructed of (concrete, concrete blocks, or gypsum wallboard); and the type of raceway penetrating the wall or floor (cable ladder, conduit, or pipe).

Firestopping detail drawings provide a section view of the fire-rated wall or floor and the raceway penetration. **See Figure 6-13.** Firestopping details also provide exact information for each particular firestopping situation. Information includes the type of floor or wall; the maximum permissible diameter of the opening; the type of raceway penetrating the wall or floor; the type of firestopping material to use; and how to apply the firestopping material.

Conduit Fill Detail Drawings. In most commercial VDV installations, cabling is not installed in conduit. Cables are instead routed in open space and supported by a cable tray, J-hangers, or other cable supports. Plenum-rated cable with a fire-resistant jacket is used in these applications. *Plenum* is a duct or area connected to distribution ducts through which air moves. Air in these ducts may be maintained at a higher pressure than normal atmospheric pressure. Plenum-rated cable does not burn or emit fumes as readily as nonplenum-rated cable, but it is more expensive.

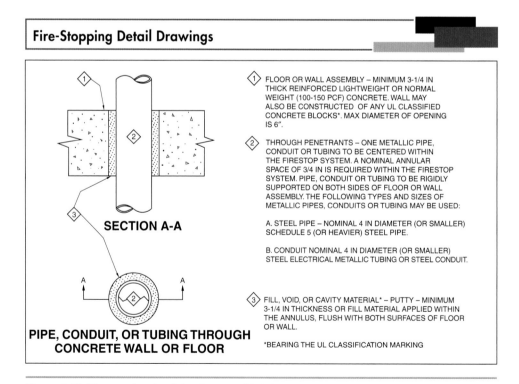

Figure 6-13. Firestopping detail drawings provide information on how to correctly apply firestopping material in walls and floors.

In some commercial VDV installations, the cable is installed in conduit. VDV cables are installed in conduit sleeves for physical protection and to allow for the ease of future modifications. A conduit fill detail drawing shows the maximum number of cables allowed in each size and type of conduit. There are separate tables for different types of conduits because the inside diameter varies from type to type. The conduit fill is limited to prevent damage to the cable during the installation process. **See Figure 6-14.**

Conduit Fill Detail Drawings

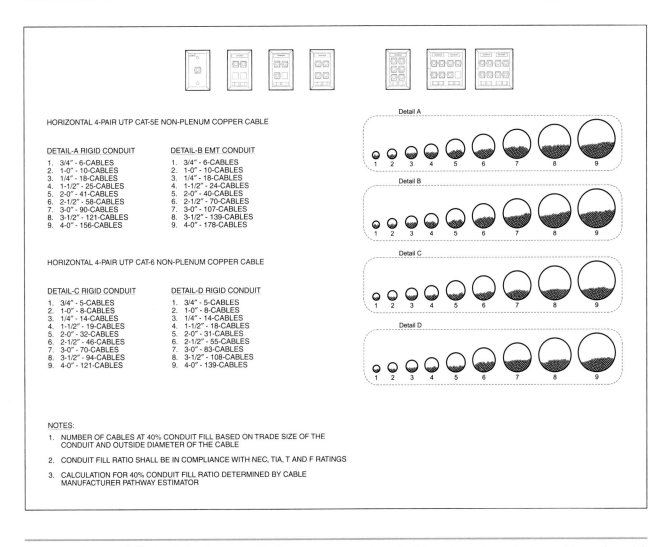

Figure 6-14. Conduit fill detail drawings provide the maximum number of cables allowed in each size and type of conduit involved in a project.

Summary

In order to properly install, expand, revise, and troubleshoot VDV equipment and systems, a VDV technician must be able to read and interpret VDV system prints and understand applicable abbreviations, symbols, and legends. VDV system prints include riser diagrams, floor plans, and detail drawings. Detail drawings include specific drawings for equipment rooms, VDV outlets, firestopping systems, and conduit fill.

Chapter Review

1. What variables determine the number and type of prints required for a specific VDV project?

2. Briefly explain why VDV systems are installed separately from most other power systems in a project.

3. What is included in a legend on a VDV print?

4. What is an outlet address?

5. What geometric shapes are used with legend number and letter references?

6. Briefly explain the purpose of VDV riser diagrams in a print.

7. What items are typically shown on a VDV floor plan?

8. What cables are included in a typical residential VDV installation? How are they routed?

9. What is a firestop?

10. Explain some differences between plenum-rated and nonplenum-rated cables.

Chapter Activity Reading VDV Prints

When installing VDV cables in conduit, a VDV technician must not exceed the conduit's fill capacity. Cable fill charts on VDV prints include information on cable types, conduit types, conduit sizes, and conduit fill capacities.

Use the communications conduit diagram to answer the following questions.

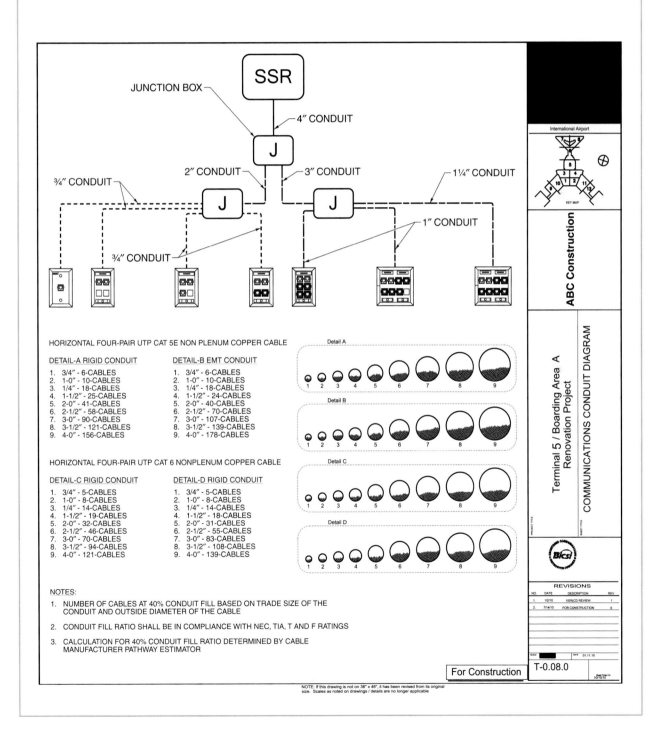

HORIZONTAL FOUR-PAIR UTP CAT 5E NON PLENUM COPPER CABLE

DETAIL-A RIGID CONDUIT
1. 3/4" - 6-CABLES
2. 1-0" - 10-CABLES
3. 1/4" - 18-CABLES
4. 1-1/2" - 25-CABLES
5. 2-0" - 41-CABLES
6. 2-1/2" - 58-CABLES
7. 3-0" - 90-CABLES
8. 3-1/2" - 121-CABLES
9. 4-0" - 156-CABLES

DETAIL-B EMT CONDUIT
1. 3/4" - 6-CABLES
2. 1-0" - 10-CABLES
3. 1/4" - 18-CABLES
4. 1-1/2" - 24-CABLES
5. 2-0" - 40-CABLES
6. 2-1/2" - 70-CABLES
7. 3-0" - 107-CABLES
8. 3-1/2" - 139-CABLES
9. 4-0" - 178-CABLES

HORIZONTAL FOUR-PAIR UTP CAT 6 NONPLENUM COPPER CABLE

DETAIL-C RIGID CONDUIT
1. 3/4" - 5-CABLES
2. 1-0" - 8-CABLES
3. 1/4" - 14-CABLES
4. 1-1/2" - 19-CABLES
5. 2-0" - 32-CABLES
6. 2-1/2" - 46-CABLES
7. 3-0" - 70-CABLES
8. 3-1/2" - 94-CABLES
9. 4-0" - 121-CABLES

DETAIL-D RIGID CONDUIT
1. 3/4" - 5-CABLES
2. 1-0" - 8-CABLES
3. 1/4" - 14-CABLES
4. 1-1/2" - 18-CABLES
5. 2-0" - 31-CABLES
6. 2-1/2" - 55-CABLES
7. 3-0" - 83-CABLES
8. 3-1/2" - 108-CABLES
9. 4-0" - 139-CABLES

NOTES:
1. NUMBER OF CABLES AT 40% CONDUIT FILL BASED ON TRADE SIZE OF THE CONDUIT AND OUTSIDE DIAMETER OF THE CABLE
2. CONDUIT FILL RATIO SHALL BE IN COMPLIANCE WITH NEC, TIA, T AND F RATINGS
3. CALCULATION FOR 40% CONDUIT FILL RATIO DETERMINED BY CABLE MANUFACTURER PATHWAY ESTIMATOR

For Construction T-0.08.0

NOTE: If this drawing is not on 36" x 48", it has been revised from its original size. Scales as noted on drawings / details are no longer applicable

Chapter Activity | Reading VDV Prints

1. Which detail provides the conduit fill capacity for four-pair UTP Cat 5e nonplenum copper cable in EMT conduit?

2. Which detail provides the conduit fill capacity for four-pair UTP Cat 6 nonplenum copper cable in rigid conduit?

3. The number of cables shown in the diagram is based on what percentage of fill?

4. How many four-pair UTP Cat 5e nonplenum copper cables are permitted in 2½″ EMT conduit?

5. How many four-pair UTP Cat 5e nonplenum copper cables are permitted in 2½″ rigid conduit?

6. Why is there a difference in the number of Cat 5e cables permitted in 2½″ EMT conduit versus 2½″ rigid conduit?

7. How many four-pair UTP Cat 6 nonplenum copper cables are permitted in 2½″ EMT conduit?

8. Why is there a difference in the number of Cat 5e versus Cat 6 cables permitted in 2½″ EMT conduit?

Chapter 7

VDV Codes and Standards

VDV installations are governed by regulatory codes. These include federal, state, and local or municipal codes. The primary code that governs VDV installations is the National Electrical Code® (NEC®). In addition to the NEC®, project specifications and ANSI/TIA standards also govern VDV installations. VDV technicians must understand and follow project specifications, ANSI/TIA standards, and the NEC® in order to correctly complete installations. Specifications, standards, and requirements can vary from job to job due to the different demands of VDV designers or engineers or due to differences in local codes.

OBJECTIVES

- Describe the CSI MasterFormat® divisions and section numbering system.
- Explain ANSI/TIA standards as they relate to VDV installations.
- Explain the National Electrical Code® and the articles related to VDV installations.
- Identify the National Electrical Code® articles that cover more than one VDV installation topic.

Digital Resources

ATPeResources.com/QuickLinks
Access Code 838502

SPECIFICATIONS

A *specification* is written information that is included with a set of prints. Specifications (also referred to as "specs") provide details related to materials and installation methods that cannot be shown on a print or that require further information. Specifications along with the various drawings describe an entire building or project, including information involving legal issues, materials, equipment placement, installation, construction, and quality. **See Figure 7-1.**

Architects, designers, and engineers develop specifications based on the requirements of the building or owner. Specifications list codes, ordinances, and company policies that must be followed for a project. For example, a note such as, "Complete VDV installation in full accordance with the *2017 National Electrical Code*® (NEC®)" may be listed on a building specification.

Municipal building departments use project specifications and prints to verify that a proposed project complies with local building codes.

Specifications

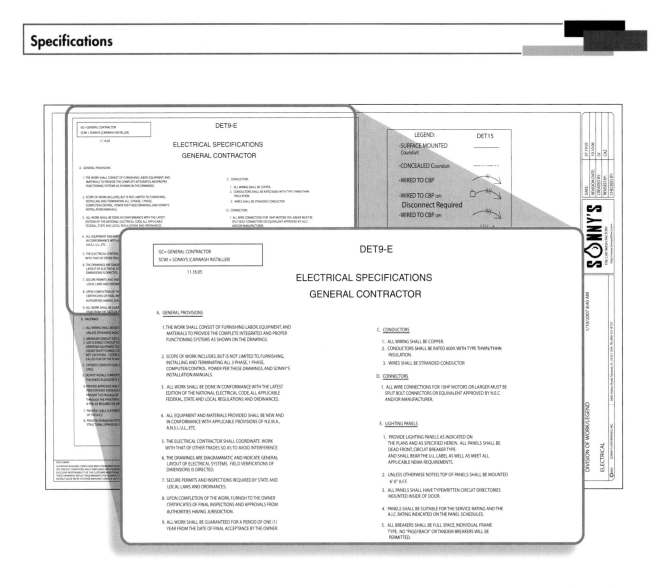

Sonny's Enterprises, Inc.

Figure 7-1. Specifications provide information on a building or project that cannot be shown on a print.

Contractors use specifications and prints to accurately bid on a project and then construct the project to meet the requirements of the owner. Project specifications may be used during subsequent inspections by a local code enforcement official, also known as the authority having jurisdiction (AHJ). An AHJ is a person who has the delegated authority to determine, mandate, and enforce code requirements established by jurisdictional governing bodies.

Specifications are intended to supplement print drawings, and the specifications must comply with the information included with the prints. When a conflict exists between the specifications and the prints, the architect, designer, or engineer must be contacted for clarification. It is common, however, for a set of specifications to stipulate that when a conflict exists between the specifications and the job-specific drawings, the drawings shall take precedence.

A large set of specifications contains many sections. If a conflict exists between different sections, then the architect, designer, or engineer must be consulted. At times, a contractor may want to deviate from the specifications, such as substituting a different brand of Cat 6 cable. The contractor must obtain permission before deviating from the specifications. Failure to follow specifications can result in non-operational systems, monetary penalties, and possible legal action. The project owner or customer may seek monetary damages in a civil court from a contractor who fails to follow specifications, and the legal proceedings can be expensive and time consuming.

The size, format, and complexity of the specifications vary with each project. For example, the specifications for a small residential project may consist of a page or two of requirements attached to the prints. The specifications for a large commercial project may consist of a 100 pages or more of detailed requirements bound together as a book. Specifications for an industrial project will typically have thousands of pages of detailed requirements.

Specifications can be used to determine installation locations and OEMs for equipment such as cable trays.

CSI MasterFormat®

The *Construction Specifications Institute (CSI)* is an organization that develops standardized construction specifications. The CSI—in cooperation with the American Institute of Architects (AIA), the Associated General Contractors of America (AGCA), the Associated Specialty Contractors (ASC), and other industry groups—has developed the MasterFormat® for construction specifications. The *MasterFormat* is a master list of numbers and titles for organizing information about construction requirements, products, and activities into a standard sequence. CSI continually promotes the MasterFormat and updates it periodically.

The MasterFormat consists of front end documents and 50 divisions. **See Figure 7-2.** The front end documents contain bidding requirements, contracting requirements, and conditions related to a construction project. The 50 divisions of the body are numbered and designed to give complete written information about individual construction requirements for building and material needs.

CSI MasterFormat® Master List

FRONT END DOCUMENTS

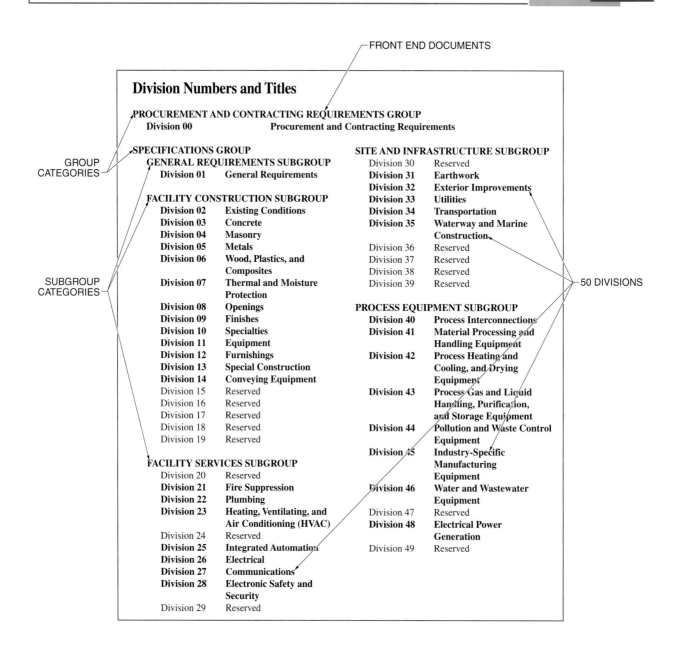

GROUP CATEGORIES

SUBGROUP CATEGORIES

50 DIVISIONS

Division Numbers and Titles

PROCUREMENT AND CONTRACTING REQUIREMENTS GROUP
Division 00 — Procurement and Contracting Requirements

SPECIFICATIONS GROUP
GENERAL REQUIREMENTS SUBGROUP
Division 01 — General Requirements

FACILITY CONSTRUCTION SUBGROUP
Division 02 — Existing Conditions
Division 03 — Concrete
Division 04 — Masonry
Division 05 — Metals
Division 06 — Wood, Plastics, and Composites
Division 07 — Thermal and Moisture Protection
Division 08 — Openings
Division 09 — Finishes
Division 10 — Specialties
Division 11 — Equipment
Division 12 — Furnishings
Division 13 — Special Construction
Division 14 — Conveying Equipment
Division 15 — Reserved
Division 16 — Reserved
Division 17 — Reserved
Division 18 — Reserved
Division 19 — Reserved

FACILITY SERVICES SUBGROUP
Division 20 — Reserved
Division 21 — Fire Suppression
Division 22 — Plumbing
Division 23 — Heating, Ventilating, and Air Conditioning (HVAC)
Division 24 — Reserved
Division 25 — Integrated Automation
Division 26 — Electrical
Division 27 — Communications
Division 28 — Electronic Safety and Security
Division 29 — Reserved

SITE AND INFRASTRUCTURE SUBGROUP
Division 30 — Reserved
Division 31 — Earthwork
Division 32 — Exterior Improvements
Division 33 — Utilities
Division 34 — Transportation
Division 35 — Waterway and Marine Construction
Division 36 — Reserved
Division 37 — Reserved
Division 38 — Reserved
Division 39 — Reserved

PROCESS EQUIPMENT SUBGROUP
Division 40 — Process Interconnections
Division 41 — Material Processing and Handling Equipment
Division 42 — Process Heating and Cooling, and Drying Equipment
Division 43 — Process Gas and Liquid Handling, Purification, and Storage Equipment
Division 44 — Pollution and Waste Control Equipment
Division 45 — Industry-Specific Manufacturing Equipment
Division 46 — Water and Wastewater Equipment
Division 47 — Reserved
Division 48 — Electrical Power Generation
Division 49 — Reserved

Figure 7-2. The MasterFormat® contains front end documents and 50 divisions that are used to organize construction information.

Division 27—Communications. MasterFormat Division 27 contains cabling, termination, pathway, and identification information for VDV/communications systems. Division 27 covers communication equipment rooms, communications cabling (backbone and horizontal), communications network equipment, and system supports and hangers. The division also contains quality-assurance requirements and known OEM trade names and product numbers for each section. **See Figure 7-3.**

MasterFormat Division 27— Communications

27 00 00	**Communications**
27 01 00	**Operation and Maintenance of Communications Systems**
27 01 10	Operation and Maintenance of Structured Cabling and Enclosures
27 01 20	Operation and Maintenance of Data Communications
27 01 30	Operation and Maintenance of Voice Communications
27 01 40	Operation and Maintenance of Audio-Video Communications
27 01 50	Operation and Maintenance of Distributed Communications and Monitoring
27 05 00	**Common Work Results for Communications**
27 05 13	Communication Services
27 05 13.13	Dialtone Service
27 05 13.23	T1 Services
27 05 13.33	DSL Services
27 05 13.43	Cable Services
27 05 13.53	Satellite Services
27 05 26	Grounding and Bonding for Communications Systems
27 05 28	Pathways for Communications Systems
27 05 29	Hangers and Supports for Communications Systems
27 05 33	Conduits and Backboxes for Communications Systems
27 05 36	Cable Trays for Communications Systems
27 05 39	Surface Raceways for Communications Systems
27 05 43	Underground Ducts and Raceways for Communication Systems
27 05 46	Utility Poles for Communication Systems
27 05 48	Vibration and Seismic Controls for Communication Systems
27 05 53	Identification for Communication Systems
27 06 00	**Schedules for Communications**
27 06 10	Schedules for Structured Cabling and Enclosures
27 06 20	Schedules for Data Communications
27 06 30	Schedules for Voice Communications
27 06 40	Schedules for Audio-Video Communications
27 06 50	Schedules for Distributed Communications and Monitoring
27 08 00	**Commissioning of Communications**

Figure 7-3. MasterFormat Division 27 contains cabling, termination, pathway, and identification information for VDV/communications systems.

Division 28 — Electronic Safety and Security. MasterFormat Division 28 contains information for a wide range of electronic safety and security systems. These systems include access control systems, video surveillance systems, intrusion detection systems, fire detection systems, and specialized detection systems, such as for radiation.

Section Numbering System

Every MasterFormat division is divided into sections. Each section has a six-digit reference number representing three levels of detail. For example, the reference number for structured cabling is 27 10 00. Additional items related to structured cabling systems have reference numbers starting with 27 11 00 and ending with 27 16 00.

Some sections may have two additional reference numbers representing a fourth level. For example, the Division 27 number 27 13 33.13, Communications Coaxial Splicing and Terminations, represents four levels. **See Figure 7-4.** The division number 27 represents level one. The level two number is 13. The level three number is 33. And when additional levels of clarification are needed, the fourth level number, which is 13 in this case, is preceded by a decimal point.

The MasterFormat also includes a keyword index of requirements, products, and activities. Not all sections of the MasterFormat will appear in every set of print specifications. Only the sections that are applicable to a construction project will appear in the specifications.

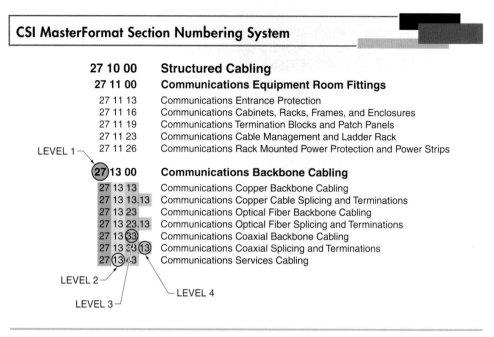

Figure 7-4. Every CSI MasterFormat division is divided into sections with each section having a six- or eight-digit reference number that represents three or four levels.

ANSI/TIA STANDARDS

The *Telecommunications Industry Association (TIA)* is an international trade group that represents several hundred telecommunications companies. The Electronic Industries Alliance (EIA) was formerly a trade association that represented US electronic equipment manufacturers. The EIA ceased operation in February 2011, but the TIA continues to serve the telecommunications industry in a similar capacity.

The TIA and EIA jointly developed standards for the telecommunications industry. The standards addressed physical infrastructure, such as cabling and connectors, and telecommunications equipment, such as telephones. The purpose of the standards was to ensure a standard physical infrastructure that would operate without regard to equipment brand.

The *American National Standards Institute (ANSI)* is a national organization that helps identify industrial and public needs for standards. Like the EIA, it was founded in the early 20th century, though the EIA was a bit more equipment focused. ANSI is much broader in scope, writing standards for more than just communications systems. The EIA and the ANSI are separate organizations,

but where their interests overlap, they work together to develop standards, such as the ANSI/TIA-586 standard used with structured cabling systems.

ANSI/TIA-568

ANSI/TIA-568 is a group of standards that cover VDV cabling for commercial buildings and other customer premises. The group consists of five individual standards: ANSI/TIA-568.0-D, ANSI/TIA-568.1-D, ANSI/TIA-568-C.2, ANSI/TIA-568.3-D, and ANSI/TIA-568-C.4.

- ANSI/TIA-568.0-D, *Generic Telecommunications Cabling for Customer Premises*, covers requirements for a variety of locations, including commercial locations such as airports.
- ANSI/TIA-568.1-D, *Commercial Building Telecommunications Infrastructure Standard*, covers general requirements.
- ANSI/TIA-568-C.2, *Balanced Twisted-Pair Telecommunications Cabling and Components Standards*, covers balanced twisted-pair cabling components. The pin-and-pair assignment for eight-conductor balanced twisted-pair cabling can be found in ANSI/TIA-568-C.2. The termination schemes are T568A

and T568B. When referencing VDV standards, tip wire designates positive polarity, and ring wire designates negative polarity. **See Figure 7-5.**

- ANSI/TIA-568.3-D, *Optical Fiber Cabling and Components Standard*, covers fiber-optic cabling and related components.
- ANSI/TIA-568-C.4, Revision C, July 11, 2011, *Broadband Coaxial Cabling and Components Standard*, covers coaxial cabling and related components.

ANSI/TIA-569

ANSI/TIA-569, *Commercial Building Standard for Telecommunications Pathways and Spaces,* provides design information for commercial buildings. Cabling routes as well as the locations and types of spaces allocated for telecommunications equipment are covered in this standard.

ANSI/TIA-606

ANSI/TIA-606, *Administration Standard for the Commercial Telecommunications Infrastructure,* provides a consistent administration protocol that is independent of equipment type or manufacturer and application. Labeling and identification of installed telecommunication infrastructure components and documentation of the infrastructure are covered in this standard.

ANSI/TIA-607

ANSI/TIA-607, *Commercial Building Grounding and Bonding Requirements for Telecommunications,* provides design information related to grounding and bonding telecommunications infrastructure in commercial buildings. Grounding installation rules are covered in this standard, including conductor sizes, the components of grounding systems, and the locations of grounding connections.

> **Tech Tip**
>
> The US Government has standardized the T568A wiring scheme. It is the required wiring scheme for all government installations and in all performance categories.

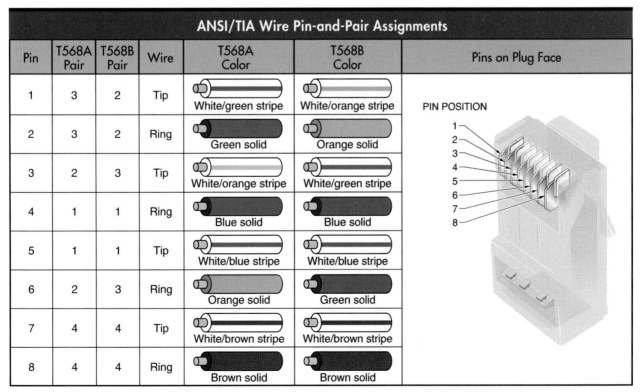

Pin	T568A Pair	T568B Pair	Wire	T568A Color	T568B Color	Pins on Plug Face
1	3	2	Tip	White/green stripe	White/orange stripe	PIN POSITION
2	3	2	Ring	Green solid	Orange solid	1 2 3 4 5 6 7 8
3	2	3	Tip	White/orange stripe	White/green stripe	
4	1	1	Ring	Blue solid	Blue solid	
5	1	1	Tip	White/blue stripe	White/blue stripe	
6	2	3	Ring	Orange solid	Green solid	
7	4	4	Tip	White/brown stripe	White/brown stripe	
8	4	4	Ring	Brown solid	Brown solid	

ANSI/TIA Wire Pin-and-Pair Assignments

Figure 7-5. The pin-and-pair assignment for eight-conductor, balanced twisted-pair cabling can be found in ANSI/TIA-568-C.2, and the termination schemes are T568A and T568B.

The purpose of the NEC® is to protect people and property from the dangers associated with the use of electricity.

THE NATIONAL ELECTRICAL CODE®

The *National Electrical Code®* (NEC®) is published by the National Fire Protection Association (NFPA), an international nonprofit organization. The NEC® is also known as the NFPA 70®. The NFPA's mission is to reduce the hazards associated with fire by providing codes, standards, and training related to fire prevention and life safety. The NEC® is updated and revised every three years to stay current with changes in technology, installation methods, and installation materials.

The NEC® covers electrical installations in residential, commercial, and industrial locations. It is a written document based on the input from electrical inspectors; OEMs of electrical components and devices; electrical industry subject matter experts; and labor, management, and trade groups representing specific industries, such as petrochemical, agricultural, mining, food processing, and heavy equipment manufacturing.

The NEC® is the primary electrical code used in the United States. Some cities, counties, and states have municipal electrical codes that are used in conjunction with the NEC®, but municipal codes contain provisions that are typically more stringent. The NEC® is not a set of laws. It becomes law only when a municipality adopts it. For this reason, it is not unusual for some municipalities to use a previous edition of the NEC® rather than the latest available edition.

The NEC® is not intended to be an instruction manual, a how-to guide, or a design guide. An electrical installation completed per the NEC® is fundamentally safe but may not allow for future expansion.

The NEC® is comprised of nine chapters in addition to the introduction and annex sections. The NEC® is arranged in a simple outline format. Within each chapter, there are sequentially numbered articles for specific topics. Each article has multiple paragraphs with sequential numbering. These paragraphs may contain lettered and numbered subparagraphs. **See Figure 7-6.** The annex section is broken into annexes A through I and contains informative material relating to topics such as product safety; conduit fill; load calculations; types of construction, administration, and enforcement; and recommended tightening torque guidelines.

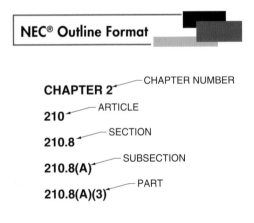

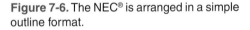

Figure 7-6. The NEC® is arranged in a simple outline format.

Frequently Referenced NEC® Articles

Although it is possible for any article of the NEC® to apply to a VDV installation, some articles are applied and referenced more frequently than others. For VDV technicians, use of these articles is a daily occurrence. Therefore, all technicians must be familiar with these articles and have access to them at all times.

Article 90—Introduction. Article 90 provides explanatory information on the NEC®. The NEC's purpose; installations that are covered; the organization of chapters and articles; the terms used; the units of measure used; mandatory rules; and planning for future expansion and convenience are covered and explained in detail in Article 90.

Article 100—Definitions. Article 100 provides definitions for terms commonly found throughout the NEC®. The terms are listed alphabetically, and only those terms which are found in more than one article of the Code are listed.

Article 110—Requirements for Electrical Installations. Article 110 provides the general requirements for all electrical installations covered by the NEC®. Approval of electrical materials, execution of work, electrical connections, marking of electrical equipment, types of electrical enclosures, spaces adjacent to electrical equipment, enclosures intended for personal entry, and tunnel installations are covered and explained in detail in Article 110.

Article 250—Grounding and Bonding. Article 250 provides the requirements for grounding and bonding of electrical systems covered by the NEC®. Systems required to be grounded or bonded; the size, type, and location of grounding and bonding conductors; grounding and bonding methods; and alternatives to grounding, such as isolation or insulation, are covered and explained in detail in Article 250.

Article 300—Wiring Methods and Materials. Article 300 provides the requirements for all wiring installations, regardless of the wiring method. Specific requirements and limitations for conductor and cable installation, raceway installation, and junction box installation, as well as locations of electrical installations are covered and explained in detail in Article 300.

Article 645—Information Technology Equipment. Article 645 provides the requirements for electrical installations in information technology equipment (computer) rooms. Definitions related to information technology equipment; equipment requirements; power wiring; interconnecting wiring; and grounding in information technology equipment rooms are covered and explained in detail in Article 645.

Article 770—Optical Fiber Cables and Raceways. Article 770 provides the requirements for fiber-optic cables and related raceways that are covered by the NEC®. Definitions; general requirements for indoor and outdoor installations; types of fiber-optic cable; and specific cable and installation requirements for different locations are covered and explained in detail in Article 770.

Article 800—Communication Circuits. Article 800 provides the requirements for telecommunications circuits and equipment that are covered by the NEC®. Definitions; general requirements for indoor and outdoor installations; grounding; types of communication cable; and specific cable and installation requirements for different locations are covered and explained in detail in Article 800.

Article 820—Community Antenna Television and Radio Distribution System. Article 820 provides the requirements for coaxial cable that distributes radio frequency (RF) signals for cable TV and camera systems. Definitions; general requirements for indoor and outdoor installations; grounding; types of coaxial cable; and specific cable and installation requirements for different locations are covered and explained in detail in Article 820.

Tech Tip

VDV technicians are encouraged to join and participate in the NFPA and NEC® development process, which helps individuals maintain awareness of the latest technology and changes in the industry. Participating in professional organizations aids in gaining knowledge about the VDV wiring process.

Common NEC® References in Multiple Articles

Common VDV installation topics appear in multiple articles of the NEC®. These topics appear multiple times because they apply to a variety of installations and because of the importance the NEC® panels place on them. Typically, the wording for common installation topics is almost identical between articles, or an article may reference another article and require it to be followed. For example, although Article 800 is titled and covers "Communications Circuits," it references other articles over 100 times.

Mechanical Execution of Work. Article 110 states, "Electrical equipment shall be installed in a neat and workmanlike manner." Similar statements regarding the mechanical execution of work are also found in Article 770, Article 800, and Article 820.

The NEC® emphasizes professional installations. Wires should be neat and orderly (neatly bundled and labeled). Raceways must be level. Switches, boxes, and connectors should be properly installed. Work areas must be properly cleaned, organized, and have adequate work space clearance after completion of work. **See Figure 7-7.**

Mechanical Execution of Work

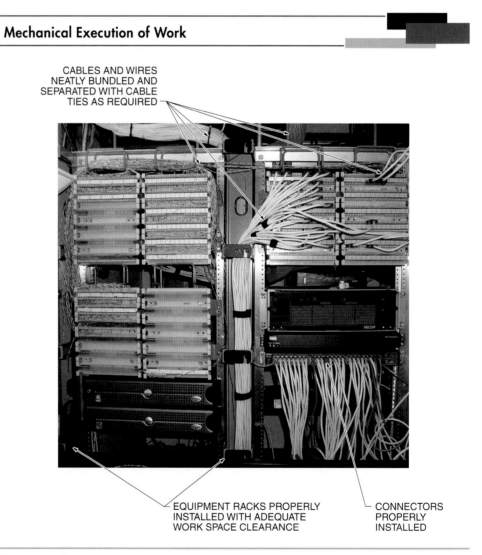

CABLES AND WIRES NEATLY BUNDLED AND SEPARATED WITH CABLE TIES AS REQUIRED

EQUIPMENT RACKS PROPERLY INSTALLED WITH ADEQUATE WORK SPACE CLEARANCE

CONNECTORS PROPERLY INSTALLED

Figure 7-7. Proper mechanical execution of work requires neat and orderly installation of equipment and devices.

Cable Supports in Suspended Ceilings. Article 300 states that cables run in suspended ceilings shall not be supported by ceiling-support wires or by conduits. Independent cable-support brackets and wires are required. Cable-support wires must be secured at each end and must be marked to distinguish them from ceiling-support wires. This article is referenced in Article 770, Article 800, and Article 820. **See Figure 7-8.**

Preventing the Spread of Fire. Article 300 states that electrical installations shall be performed so that the spread of fire will not be increased. It also states that penetrations in fire-rated assemblies, ceilings, floors, or walls shall be sealed with approved firestopping material, such as sheet-metal chase devices or fire-rated caulking compound. Similar statements are also found in Article 770, Article 800, and Article 820.

Cable Supports in Suspended Ceilings

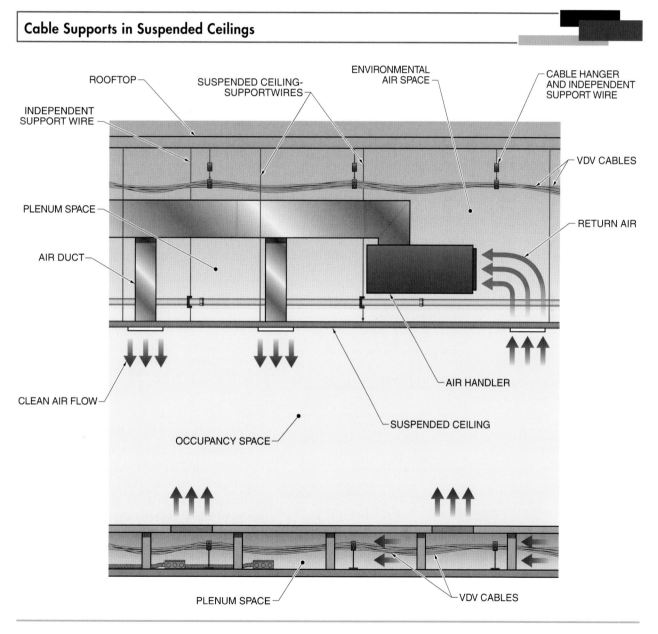

Figure 7-8. Independent cable-support brackets and wires are required when installing cables that run through spaces above suspended ceilings.

The primary mission of the NFPA, the publisher of the *National Electrical Code®*, is to reduce the hazards associated with fire and prevent damage to property and injuries to personnel. **See Figure 7-9.**

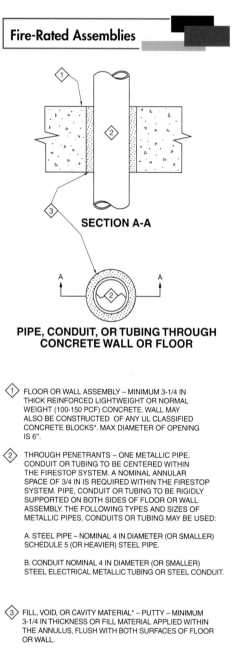

Fire-Rated Assemblies

SECTION A-A

PIPE, CONDUIT, OR TUBING THROUGH CONCRETE WALL OR FLOOR

1. FLOOR OR WALL ASSEMBLY – MINIMUM 3-1/4 IN THICK REINFORCED LIGHTWEIGHT OR NORMAL WEIGHT (100-150 PCF) CONCRETE. WALL MAY ALSO BE CONSTRUCTED OF ANY UL CLASSIFIED CONCRETE BLOCKS*. MAX DIAMETER OF OPENING IS 6".

2. THROUGH PENETRANTS – ONE METALLIC PIPE, CONDUIT OR TUBING TO BE CENTERED WITHIN THE FIRESTOP SYSTEM. A NOMINAL ANNULAR SPACE OF 3/4 IN IS REQUIRED WITHIN THE FIRESTOP SYSTEM. PIPE, CONDUIT OR TUBING TO BE RIGIDLY SUPPORTED ON BOTH SIDES OF FLOOR OR WALL ASSEMBLY. THE FOLLOWING TYPES AND SIZES OF METALLIC PIPES, CONDUITS OR TUBING MAY BE USED:

   A. STEEL PIPE – NOMINAL 4 IN DIAMETER (OR SMALLER) SCHEDULE 5 (OR HEAVIER) STEEL PIPE.

   B. CONDUIT NOMINAL 4 IN DIAMETER (OR SMALLER) STEEL ELECTRICAL METALLIC TUBING OR STEEL CONDUIT.

3. FILL, VOID, OR CAVITY MATERIAL* – PUTTY – MINIMUM 3-1/4 IN THICKNESS OR FILL MATERIAL APPLIED WITHIN THE ANNULUS, FLUSH WITH BOTH SURFACES OF FLOOR OR WALL.

   *BEARING THE UL CLASSIFICATION MARKING

Figure 7-9. Statements regarding the prevention of the spread of fire are found in VDV prints.

Wiring in Air-Handling Spaces (Plenums). Frequently, the area above a suspended ceiling is used as a return plenum for an air-handling system. Article 300 states which wiring methods may be used in these air-handling spaces. The article also states that cables may only be used if they are listed for use in an air-handling space (plenum). In addition to similar statements regarding wiring in plenums, Article 770, Article 800, and Article 820 list specific cable types that are permissible in air-handling spaces.

Firestopping material is available in multiple forms for sealing penetrations, such as holes in walls, structural members, and floors used for routing cables.

Panels Designed to Allow Access. Article 300 states, "Cables, raceways, and equipment installed behind panels designed to allow access, including suspended ceiling panels, shall be arranged and secured so as to allow the removal of panels and access to the equipment." Similar statements regarding panels designed to allow access to equipment are also found in Article 770, Article 800, and Article 820. The NEC's® intent is to ensure that panels can be removed to allow easy access to equipment for maintenance or future installations. **See Figure 7-10.**

Removable Ceiling Panels

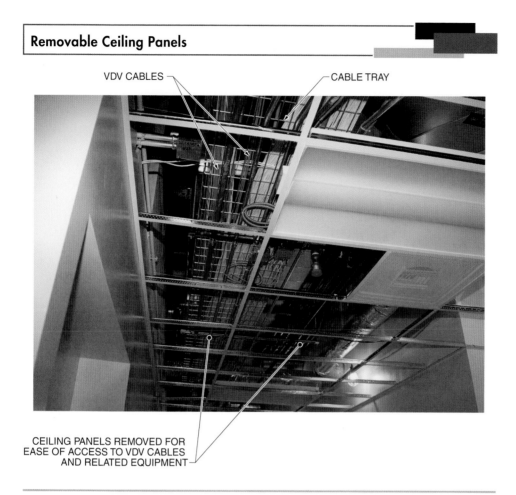

VDV CABLES

CABLE TRAY

CEILING PANELS REMOVED FOR
EASE OF ACCESS TO VDV CABLES
AND RELATED EQUIPMENT

Figure 7-10. Statements regarding panels designed to allow access to equipment are found in NEC® Article 300, Article 770, Article 800, and Article 820.

Abandoned Cables. Article 645 defines abandoned cables as cables that are not terminated at equipment and have not been marked for future use with a tag. The article also states that the accessible portions of abandoned cables that are not in a raceway must be removed. Similar statements regarding abandoned cables are also found in Article 770, Article 800, and Article 820.

Abandoned cables add to the amount of combustible products available in the event of a fire. They can also block the flow of air through air-handling spaces and under raised floors in computer rooms. The intent of the NEC® is to reduce the amount of combustible products and to provide for the free flow of air. **See Figure 7-11.**

Abandoned Cables

Figure 7-11. NEC® Article 645 defines abandoned cables as cables that are not terminated at equipment and have not been marked for future use with a tag.

Summary

VDV technicians use CSI MasterFormat, ANSI/TIA standards, and the NEC® to guide them when installing low-voltage systems. An AHJ uses the NEC® to ensure a safe installation. ANSI/TIA standards and CSI Master-Format are typically used by building owners, customers, VDV designers, or VDV engineers to further define installation requirements.

Chapter Review

1. What is the primary code that governs VDV installations?

2. What is an authority having jurisdiction?

3. What is the Construction Specifications Institute (CSI)?

4. How many divisions does MasterFormat consist of?

5. What two organizations are presently referenced instead of the Electronics Industry Alliance (EIA)?

6. What are the five individual standards that comprise ANSI/TIA-568?

7. What is another name for the National Electrical Code®?

8. Which article of the NEC® covers definitions?

9. Which article of the NEC® covers fiber-optic cables?

10. What does Article 645 of the NEC® state regarding abandoned cables that are not in a raceway?

Chapter Activity | Twisted-Pair Conductor Termination Patterns

The termination schemes for eight-conductor twisted-pair cable are T568A and T568B. *Identify the color of the conductor terminated on each pin of the RJ-45 connector.*

T568A

_____ **1.** Pin 1

_____ **2.** Pin 2

_____ **3.** Pin 3

_____ **4.** Pin 4

_____ **5.** Pin 5

_____ **6.** Pin 6

_____ **7.** Pin 7

_____ **8.** Pin 8

T568B

_____ **9.** Pin 1

_____ **10.** Pin 2

_____ **11.** Pin 3

_____ **12.** Pin 4

_____ **13.** Pin 5

_____ **14.** Pin 6

_____ **15.** Pin 7

_____ **16.** Pin 8

Chapter 8

Grounding and Bonding VDV Systems

Grounding and bonding are required for residential, commercial, and industrial power and lighting installations operating at voltages of 120 VAC or higher. Although most VDV systems operate at much lower voltages—typically less than 50 V—grounding and bonding are also required for these systems. VDV systems must be grounded and bonded to protect personnel from harm and VDV system equipment from damage. Grounding and bonding also minimize electrical noise and ensure proper performance of sensitive electronic components that are part of a VDV system, such as servers, routers, and network switches.

OBJECTIVES

- Explain the purpose of grounding and bonding VDV systems.
- Compare grounding and bonding electrical equipment and systems and grounding and bonding VDV equipment and systems.
- Describe an intersystem bonding termination.
- Summarize the different requirements of the NEC® and TIA.
- Describe different grounding methods.
- Describe different grounding and bonding conductors and devices.

Digital Resources

ATPeResources.com/QuickLinks
Access Code 838502

GROUNDING AND BONDING PRINCIPLES

Grounding is a low-resistance conducting connection between electrical circuits, equipment, and the earth. When discussing grounding and bonding, the terms "ground" and "earth" may be used interchangeably. Outside the United States, ground may be referred to as "earth," and the act of grounding may be referred to as "earthing."

A *grounding electrode* is a conductive metal object used to establish a connection from an electrical circuit to the earth. A *grounding electrode conductor (GEC)* is a conductor that connects parts of an electrical distribution system (equipment grounding conductors, grounded conductors, and all metal parts) to an approved grounding electrode system.

Bonding is the act of connecting two metallic parts to form a continuous, low-impedance electrical path. The metallic parts may be identical or different types. Metallic parts that are bonded include conduit, equipment racks, equipment frames, cable shields, and conductors. A *bonding conductor*, or a bonding jumper, is a conductor used to bond metal objects when required. An intersystem bonding termination is a device that provides a means of connecting bonding conductors from communications systems to a grounding electrode system. This device can be a terminal bar or a copper plate. **See Figure 8-1.**

The metal components of a VDV system must be bonded and grounded. The potential of the earth is considered to be zero.

Grounding and Bonding VDV Systems

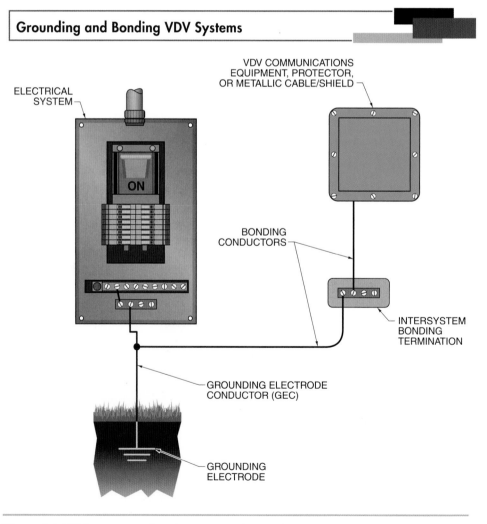

Figure 8-1. VDV systems and equipment must be bonded and grounded.

Proper grounding and bonding ensure that there is no difference of potential between the metal components of a VDV system. If the metal components of a VDV system were not grounded and bonded, a difference of potential between components could exist, which can result in the flow of unwanted current. A large difference of potential can be hazardous to personnel and equipment. A small difference of potential can result in a small unwanted current flow, which can cause sensitive electronic equipment to malfunction. **See Figure 8-2.**

Difference of Potential

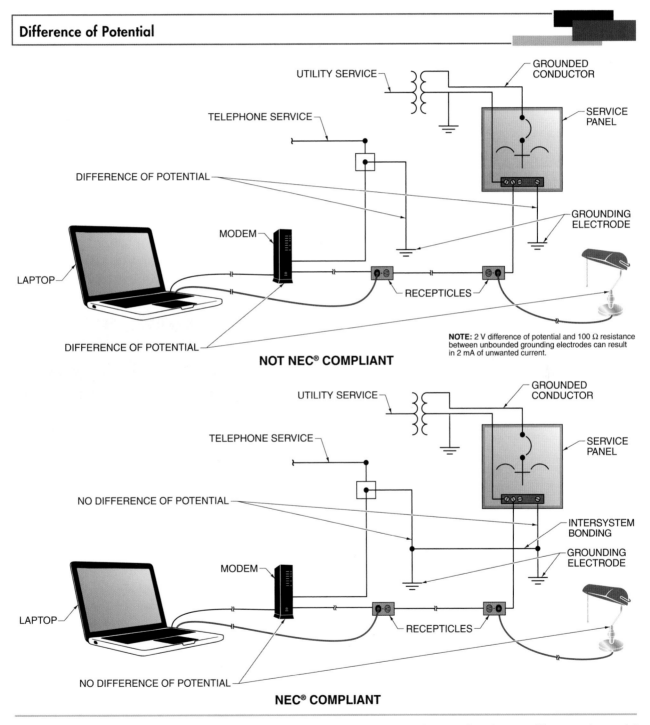

NOT NEC® COMPLIANT

NOTE: 2 V difference of potential and 100 Ω resistance between unbounded grounding electrodes can result in 2 mA of unwanted current.

NEC® COMPLIANT

Figure 8-2. Failure to effectively bond grounding electrodes can result in unwanted current flow due to a difference of potential between the electrodes.

Proper grounding and bonding protect personnel and limit damage to VDV systems by providing a safe electrical path to ground for unwanted voltages. Unwanted voltages may be caused by a lightning strike that hits part of a VDV system, accidental contact between a higher voltage source and part of a VDV system, or by some other source of a transient voltage imposed or induced on part of a VDV system. A *transient voltage* is a temporary, unwanted voltage in an electrical circuit. The transient voltage and current are safely carried along the low-impedance path of the metallic grounding system and safely dissipated to the earth. **See Figure 8-3.**

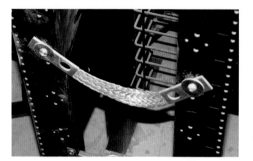

VDV racks are connected with bonding straps as part of a grounding system.

NEC® GROUNDING AND BONDING REQUIREMENTS

The NEC® contains provisions for grounding and bonding that are necessary for a safe installation. Article 250 of the NEC®, *Grounding and Bonding*, is the primary article that covers this topic. The majority of the information in Article 250 pertains to the grounding of electrical power systems. Section 250.94 of the NEC®, *Bonding for Other Systems*, covers the bonding of other systems to the grounding electrode system for a building electrical service. This is known as intersystem bonding. A *grounding electrode system* is a system of all the grounding conductors in a building or structure that are bonded together.

Specific information on the grounding and bonding of VDV systems is found in Article 770, *Optical Fiber Cables and Raceways;* Article 800, *Communication Circuits;* and Article 820, *Community Antenna Television and Radio Distribution Systems.* Part III of each article is titled *Protection,* and Part IV of each article is titled *Grounding Methods.*

Grounding and Bonding VDV Systems

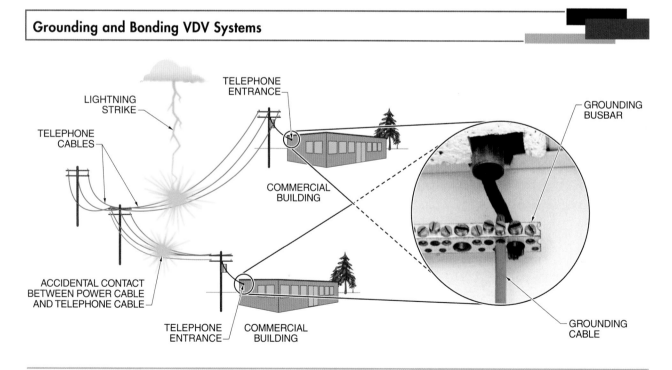

Figure 8-3. Proper grounding and bonding provides a low-impedance path to ground for transient voltages.

Intersystem Bonding Terminations

An *intersystem bonding termination* is a device that provides a means of connecting bonding conductors from communications systems to a grounding electrode system. This device can be a terminal bar or a copper plate. Often, VDV systems are installed or connected after a building's electrical power system is installed and energized. An intersystem bonding termination provides a safe and convenient location for connecting VDV bonding conductors without the need to access the building's electrical power system. **See Figure 8-4.** There are several requirements for an intersystem bonding termination:

- It must be accessible and not interfere with the operation or the opening or closing of the equipment it is mounted on or adjacent to.

- It must provide a minimum of three terminals for intersystem bonding conductors.

- It must be mounted and electrically connected to the service enclosure, the enclosure for building disconnecting means, the meter enclosure, or the metallic power service conduit (nonflexible).

- Alternatively, it may be mounted at the service enclosure, the enclosure for building disconnecting means, or the meter enclosure and connected to the metallic enclosure or the grounding electrode conductor with a minimum 6 AWG copper conductor.

Tech Tip

Per the NEC®, exposed grounding conductors must be covered with green- or green-and-yellow striped insulation. Green tape can be used in place of insulation on bare conductors.

Intersystem Bonding Terminations

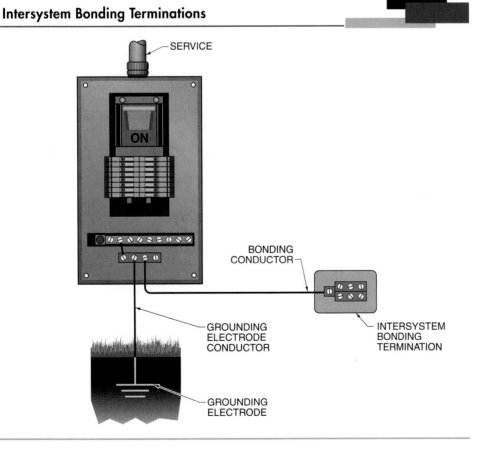

Figure 8-4. A variety of intersystem bonding termination devices are available for connecting VDV bonding conductors.

Communications Systems Grounding

Part III of Article 800, *Protection*, covers the use and installation of primary protectors for entrance communication cables. Entrance communication cables are cables which are routed from the communication service provider, or exterior of the building, to the end user, or interior of the building. Part IV of Article 800, *Grounding Methods*, covers the grounding and bonding of the primary protector and the entrance communication cables.

Primary Protectors. A *primary protector* is a surge protector that is specifically designed to protect communication cables and the equipment connected to the cables from electrical surges. Part III of Article 800 requires primary protectors for overhead (aerial) and underground communication cables that enter a building or structure. In many instances, the primary protector serves as a termination point connecting the entrance cable with the building cables. Primary protectors are located as close as possible to the cables' point of entrance. **See Figure 8-5.**

Tech Tip

Most primary protectors used in VDV systems are fuseless. If conditions for fuseless primary protectors are not met, fused primary protectors are available and consist of an arrester connected between each line conductor and ground.

Primary Protectors

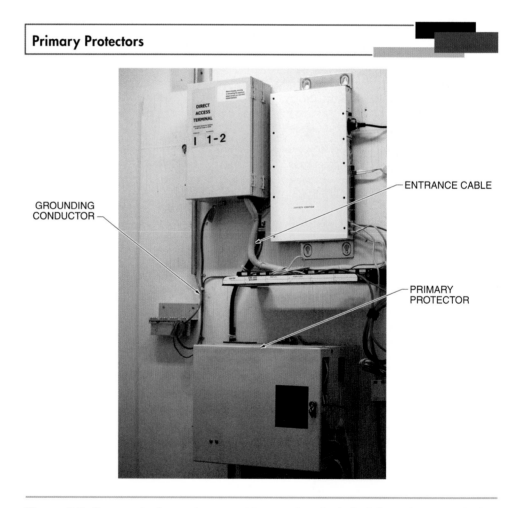

GROUNDING CONDUCTOR

ENTRANCE CABLE

PRIMARY PROTECTOR

Figure 8-5. Communication entrance cables are terminated at the primary protector. The primary protector is connected to the grounding busbar.

A *surge* is a type of transient voltage or current that usually rises rapidly to a peak value and then falls more slowly to zero. Surges are usually no longer than a few milliseconds. Surges can be caused by lightning or direct contact with a power line. Surges can also be caused by a field induced between either an entrance communication cable and lightning or an entrance communication cable and power line. If a surge is present on an entrance communication cable, personnel can be injured and sensitive electronic equipment can be damaged. Primary protectors provide overvoltage and overcurrent protection by diverting surge energy away from personnel and sensitive equipment to ground. Primary protectors use fuses or solid-state devices to handle surges.

Grounding Methods. Part IV of Article 800, *Grounding Methods*, covers the grounding and bonding of primary protectors and the metallic shields of entrance communication cables. The primary protector and the metallic shield of an entrance communication cable must be connected to a grounding electrode by a bonding conductor or a GEC. Part IV of Article 800 is divided into four major sections. The first section covers the bonding conductor or GEC. Specific requirements of this section include the following:

- The conductor must be listed. (*Note*: Materials that are listed have been inspected and tested by an organization, such as Underwriters Laboratories Inc.®, and determined to be suitable for a specific application.)
- Conductors can be insulated, covered, or bare.
- The conductor can be solid or stranded, but it must be made of copper or another corrosion-resistant conductive material.
- The minimum size of the conductor must be 14 AWG, and the maximum size must be 6 AWG.
- The conductor must have a current-carrying capacity no less than that of the metallic shield and protected conductor of the entrance communication cable.
- The length of the conductor must be as short as possible.

- In one- and two-family dwellings, the length of the conductor must not exceed 20′, but exceptions to this are specified.
- The conductor must be run as straight as possible.
- Where exposed to physical damage, the conductor must be protected.
- If the conductor is installed in a metal raceway for protection, then it must be bonded to the raceway at both ends.

The second section of Part IV of Article 800 covers the connection of the bonding conductor or GEC to the grounding electrode. There are three specific requirements in this section:

- If an intersystem bonding termination is present, then the bonding conductor must be connected to it.
- If an intersystem bonding termination is not present and a grounding means is available, the conductor must be connected to one of seven points at the nearest accessible location. **See Figure 8-6.** The seven points include the following:
  - The grounding electrode system for the building electrical service
  - A grounded interior metal water pipe within 5′ of the cable's point of entrance to the building
  - An accessible means external to the power service, as specified in Section 250.94, *Bonding for Other Systems*
  - A metallic power service conduit (nonflexible)
  - A service equipment enclosure
  - A GEC or the metal enclosure of the GEC
  - A GEC or grounding electrode for a separate building as specified in Section 250.32, *Buildings or Structures Supplied by a Feeder(s) or Branch Circuit(s)*
- If an intersystem bonding termination is not present and a grounding means is not available, there are two options for connecting a bonding conductor or GEC:

- The conductor can be connected to any grounding electrode specified in Sections 250.52(A) (1) through 250.52(A) (4), *Electrodes Permitted for Grounding*.
- If a grounding electrode as specified above is not available, an electrode complying with 250.52(A) (7) and 250.52(A) (8) can be used. These sections also permit the use of a ground rod with a minimum length of 5′ and a minimum diameter of ½″.

- The ground rod must be a minimum of 6′ from other electrodes.

The third section covers specific methods for connecting a bonding conductor or GEC to a grounding electrode. The section refers to the methods specified in Section 250.70, *Methods of Grounding and Bonding Conductor Connection to Electrodes*. Exothermic welding, listed lugs, listed pressure connectors, listed clamps, or other listed methods are all acceptable means of connection per Section 250.70. **See Figure 8-7.**

Grounding and Bonding Connection Points

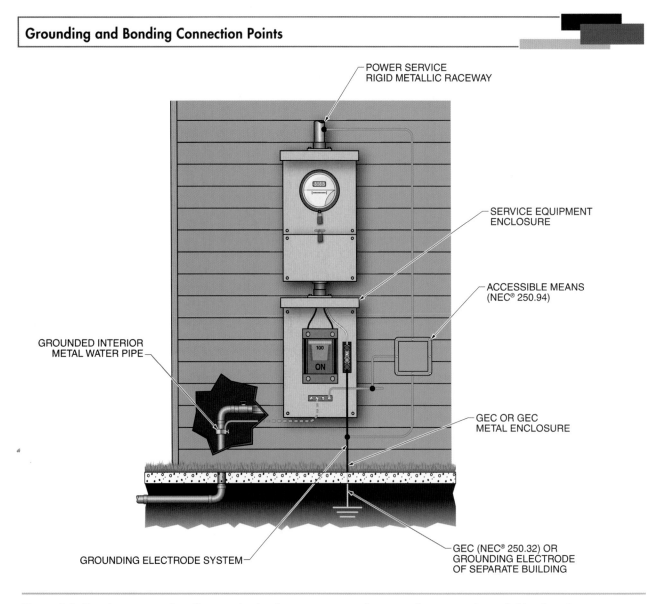

POWER SERVICE
RIGID METALLIC RACEWAY

SERVICE EQUIPMENT
ENCLOSURE

ACCESSIBLE MEANS
(NEC® 250.94)

GROUNDED INTERIOR
METAL WATER PIPE

GEC OR GEC
METAL ENCLOSURE

GROUNDING ELECTRODE SYSTEM

GEC (NEC® 250.32) OR
GROUNDING ELECTRODE
OF SEPARATE BUILDING

Figure 8-6. If an intersystem bonding termination is not present and a grounding means is available, then there are seven possible locations to connect a GEC.

Grounding and Bonding Connection Methods

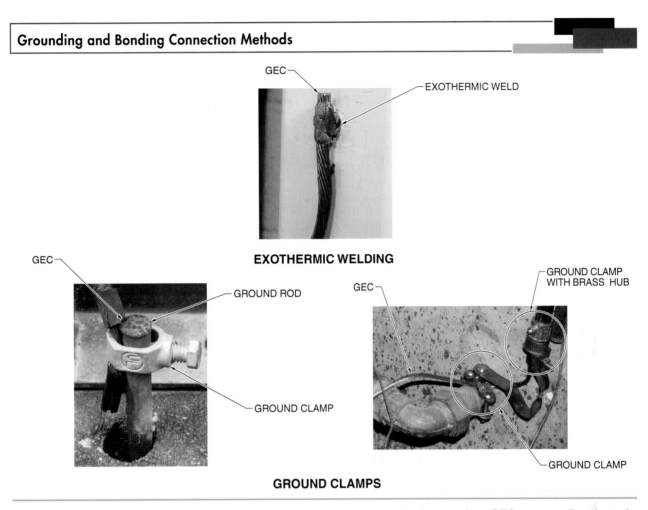

EXOTHERMIC WELDING

GROUND CLAMPS

Figure 8-7. Listed ground clamps and exothermic welding are acceptable methods of connecting a GEC to a grounding electrode.

The fourth section covers the bonding of grounding electrodes. In the event that there is a separate grounding electrode installed for the VDV system, it must be bonded to the grounding electrode system for the building electrical service. The minimum size of the bonding conductor must be 6 AWG copper. The electrodes must be bonded so there will be no difference of potential between them. **See Figure 8-8.**

Tech Tip

The shields of armored twisted-pair cables must be grounded. A shielded connector with a threaded stud provides a connection point for a bonding conductor. Shielded connectors are commonly referred to as bullet bonds.

VDV Grounding Electrodes

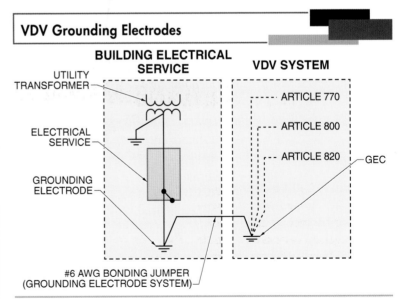

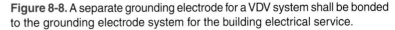

Figure 8-8. A separate grounding electrode for a VDV system shall be bonded to the grounding electrode system for the building electrical service.

TIA GROUNDING AND BONDING REQUIREMENTS

Information technology (IT) equipment evolved rapidly in the 1980s, and it became evident that the grounding requirements of the NEC® were not adequate for more sensitive electronic equipment. The NEC® primarily addressed the safety aspects of grounding. The Telecommunications Industry Association (TIA) and the Electronic Industries Alliance (EIA) initially published a standard for grounding and bonding of telecommunication systems in 1994, titled TIA/EIA-607, *Commercial Building Grounding and Bonding Requirements of Telecommunications*.

TIA/EIA-607 was not intended to replace the NEC®. TIA/EIA-607 addressed the design and installation of a generic grounding and bonding system for telecommunications that could use IT components from any vendor. TIA/EIA-607 was incorporated into projects via project specifications.

The most current standard, published in 2015, is ANSI/TIA-607-C, *Generic Telecommunications Bonding and Grounding (Earthing) for Customer Premises*. The primary components of ANSI/TIA-607-C are the following:

- primary bonding busbar (PBB)
- telecommunications bonding conductor (TBC)
- secondary bonding busbar (SBB)
- telecommunications bonding backbone (TBB)
- backbone bonding conductor (BBC)
- telecommunication equipment bonding conductor (TEBC)
- alternating current equipment ground (ACEG)

The grounding and bonding begins at a telecommunications entrance facility (TEF), the location where telecommunication cables enter a building. The number and type of components used in a building depend on the size and design of the building. ANSI/TIA-607-C does not require short metal pathways, such as floor sleeves or metal cable supports, to be bonded. **See Figure 8-9.**

Primary Bonding BusBars (PBBs)

A *primary bonding busbar (PBB)* is a busbar that serves as the central connection point between the grounding electrode system for a building electrical service and the grounding and bonding system for telecommunications. A PBB is connected to the grounding electrode system for a building electrical service by the TBC. PBBs are listed copper or copper alloy busbars that are ¼″ thick by 4″ wide and are available in various lengths. A PBB has drilled holes for mounting and for terminating listed lugs.

A PBB is located in the TEF. If the TEF has an electrical panel that serves IT loads, the PBB must be mounted near it. If the TEF does not have an electrical panel that serves IT loads, the PBB must be mounted near the terminations for backbone cabling. A PBB must be insulated from its support by a listed insulator. The recommended minimum stand-off size for the PBB insulator is 2″. This allows access to the rear of the PBB.

The connections to a PBB shall utilize exothermic welds, exothermic two-hole lugs, or listed two-hole compression lugs. The surface of a PBB must be clean and an antioxidant compound applied before lugs are attached. The following components must be connected to a PBB:

- a telecommunications bonding conductor (TBC)
- a telecommunications bonding backbone (TBB)
- building steel, if accessible in the TEF, with a bonding conductor at least 6 AWG in size
- an alternating current equipment ground (ACEG)
- a telecommunication equipment bonding conductor (TEBC)
- telecommunication equipment in the TEF
- a primary protector
- metallic shields of telecommunication cables that terminate in the TEF
- metal raceways with telecommunication cables that terminate in the TEF
- a metal cable tray with telecommunication cables in the TEF, with a bonding conductor at least 6 AWG in size (*Note*: A bonding jumper is required at the connection point for each section of the cable tray.)

ANSI/TIA-607-C, General Telecommunications Bonding and Grounding (Earthing) for Consumer Premises

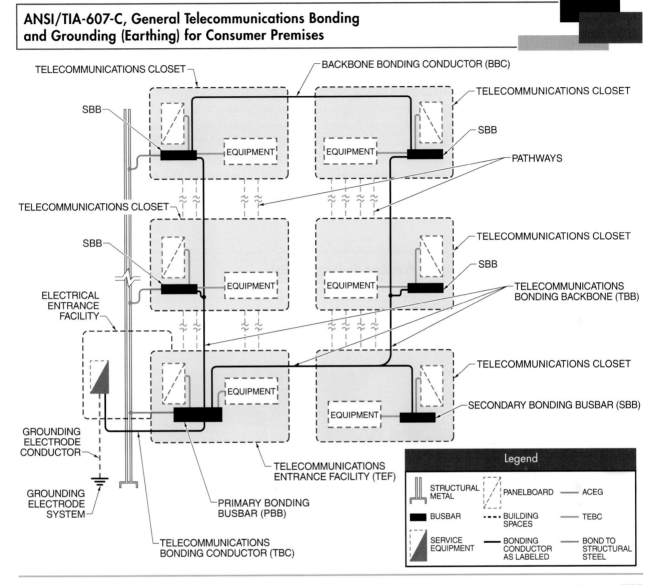

Figure 8-9. A large commercial building can have multiple telecommunications rooms and equipment rooms in addition to a TEF.

Telecommunications Bonding Conductors (TBCs)

A *telecommunications bonding conductor (TBC)* is a conductor that bonds a primary bonding busbar to the grounding electrode system for a building electrical service. This ensures that there is no difference of potential between the two ground systems. The length of the TBC should be as short as possible. Although not covered in ANSI/TIA-607-C, it is advisable to have a licensed electrician install a TBC since it connects to the building electrical service. Additional installation requirements include the following:

- A TBC must be copper and may be insulated.
- A TBC must be green or marked with a distinctive green color.
- The minimum size for a TBC must be the same size as the largest TBB.
- A TBC must be uniform in size and not decrease in size as the grounding path approaches the earth.
- A TBC should not be installed in a ferrous metal conduit.
- If it is necessary to install a TBC in a ferrous metal conduit, a bonding bushing must be used at each end to bond it to the conduit with a minimum 6 AWG conductor.

Secondary Bonding Busbars (SBBs)

A *secondary bonding busbar (SBB)* is a busbar located in the telecommunication room or equipment room of a building and connects to the primary bonding busbar. **See Figure 8-10.** Some buildings have telecommunication rooms or equipment rooms in addition to a TEF. ANSI/TIA-607-C requires that each of these rooms have an SBB. An SBB is connected to a PBB by the TBB. SBBs are listed copper or copper alloy busbars that are ¼″ thick by 2″ wide and are available in various lengths. An SBB has drilled holes for mounting and for terminating listed lugs. If a telecommunication room or equipment room has an electrical panel that serves IT loads, the SBB must be mounted near it. If a telecommunication room or equipment room does not have an electrical panel that serves IT loads, the SBB must be mounted near the terminations for backbone cabling. An SBB must be insulated from its support by a listed insulator. The recommended minimum size for the insulator is 2″ height. This allows access to the rear of the SBB.

Secondary Bonding Busbars (SBBs)

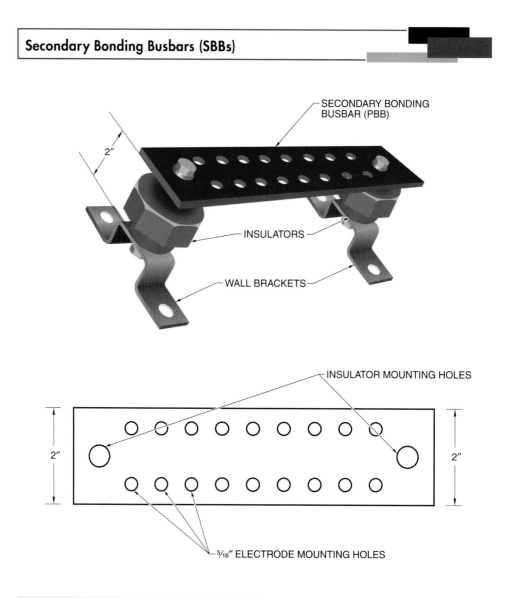

Figure 8-10. A secondary bonding busbar (SBB) must be located in a TEF per ANSI/TIA-607-C.

The connections to an SBB must utilize exothermic welds, exothermic two-hole lugs, or listed two-hole compression lugs. **See Figure 8-11.** The surface of an SBB shall be clean and an antioxidant compound applied before lugs are attached. The following components shall be connected to a SBB:

- a telecommunications bonding backbone (TBB)
- if required, a bonding jumper between the TBB and SBB that is continuous, in the straightest path possible, and as short as possible
- if accessible in the TEF, building steel with a bonding conductor that is at least 6 AWG
- an alternating current equipment ground (ACEG)
- a telecommunications equipment bonding conductor (TEBC)
- telecommunication equipment located in the telecommunication room or equipment room
- metallic shields of telecommunication cables that terminate in the telecommunication room or equipment room
- metal raceways with telecommunication cables that terminate in the telecommunication room or equipment room
- metal cable trays with telecommunication cables in the telecommunication room or equipment room, with a bonding conductor that is at least 6 AWG (*Note*: A bonding jumper is required at the connection point for each section of the cable tray.)
- a backbone bonding conductor (BBC)
- additional SBBs located in the same telecommunication room or equipment room, with a bonding conductor that is the same size as the TBB

Busbar Connection Devices

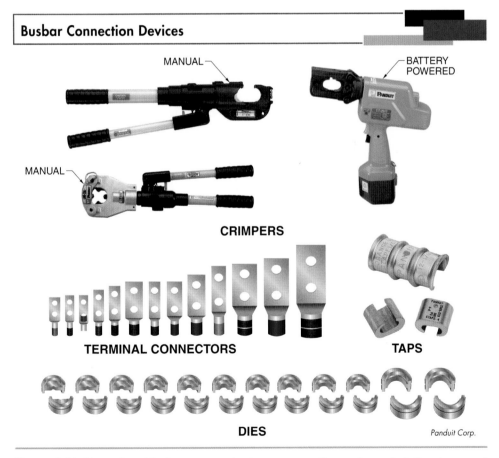

Figure 8-11. Manual and battery-powered crimpers as well as color-coded dies, lugs, and terminal connectors are used to terminate conductors to PBBs and SBBs.

Telecommunications Bonding Backbones (TBBs)

A *telecommunications bonding backbone (TBB)* is a cable that connects a primary bonding busbar to secondary bonding busbars located in telecommunication rooms or equipment rooms located throughout a building. The minimum size of a TBB is 6 AWG, and the maximum size is 750 kcmil. The required size can be found in Table 1 in ANSI/TIA-607-C. Additional installation requirements include the following:

- A TBB must be copper and may be insulated.

- A TBB must be green or marked with a distinctive green color.

- A TBB must be uniform in size and not decrease in size as the grounding path approaches the earth.

- A TBB should not be installed in a ferrous metal conduit.

- If it is necessary to install a TBB in a ferrous metal conduit, a bonding bushing must be used at each end to bond it to the conduit with a minimum 6 AWG conductor.

- A TBB shall follow the pathways of the telecommunication backbone cables.

- A TBB must be protected from damage.

- Multiple TBBs are allowed to conform to a building design.

- The metallic shield of telecommunication backbone cables or the water piping system must not be used as a TBB.

- A TBB should be installed without any splices.

- If it is necessary to splice a TBB, the following requirements must be followed:
    - The number of splices must be kept to a minimum.
    - The splices must be accessible and located in a telecommunications equipment room.
    - The splices must be supported and protected.
    - The splices must be made with an exothermic weld or a listed irreversible compressor-type connector.

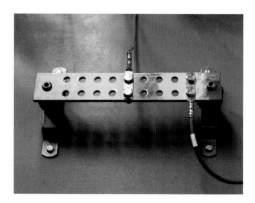

A telecommunication system is bonded through connection to a primary or secondary bonding busbar.

Backbone Bonding Conductors (BBCs)

A *backbone bonding conductor (BBC)* is a conductor that connects multiple telecommunications bonding backbone conductors. In large, multistory buildings, there may be multiple telecommunications rooms and/or equipment rooms on the same floor. The SBBs in these rooms need to be connected to ensure that there are no differences in potential between them. Backbone bonding conductors also connect multiple secondary bonding busbars on the same floor of a building. ANSI/TIA-607-C requires TBBs to be bonded at the top floor of a building with a BBC and at every third floor down to the lowest floor. The minimum size of a BBC is 6 AWG, and the maximum size is 750 kcmil. The required size can be found in Table 1 in ANSI/TIA-607-C. **See Figure 8-12.** Additional installation requirements include the following:

- A BBC must be copper and may be insulated.

- A BBC must be green or marked with a distinctive green color.

- The minimum size for a BBC must be the same size as the largest TBB.

- A BBC must be uniform in size and not decrease in size as the grounding path approaches the earth.

- A BBC should not be installed in a ferrous metal conduit.

- If it is necessary to install a BBC in a ferrous metal conduit, a bonding bushing shall be used at each end to bond it to the conduit with a minimum 6 AWG conductor.

TBB and BBC Miniumum Size Requirements	
TBB/BCC Linear Length*	Conductor Size†
Less than 4 (13)	6
4–6 (14–20)	4
7–8 (21–26)	3
9–10 (27–33)	2
11–13 (34–41)	1
14–16 (42–52)	1/0
17–20 (53–66)	2/0
21–26 (67–84)	3/0
27–32 (85–105)	4/0
33–38 (106–125)	250 kcmil
39–46 (126–150)	300 kcmil
47–53 (151–175)	350 kcmil
54–76 (176–250)	500 kcmil
77–91 (251–300)	600 kcmil
Greater than 91 (301)	750 kcmil

* m (ft)
† AWG

Figure 8-12. Table 1 of ANSI/TIA-607-C is used to determine the size of a TBB and a BBC.

Telecommunications Equipment Bonding Conductors (TEBCs)

A *telecommunications equipment bonding conductor (TEBC)* is a conductor that connects a primary bonding busbar or secondary bonding busbar to equipment cabinets or racks. There may be more than one TEBC. Rack bonding conductors (RBCs) are connected to a TEBC by listed irreversible compression lugs. A TBC bonds the PBB to the grounding electrode system for the building electrical service, and a TEBC bonds the PBB or SBB to equipment cabinets or racks. It is common to have multiple connection points along the length of a TEBC. **See Figure 8-13.** Additional installation requirements include the following:

- A TEBC must be continuous.
- A TEBC must be copper and may be insulated.
- A TEBC must be green or marked with a distinctive green color.

- The minimum size of a TEBC must be 6 AWG or the largest size equipment grounding conductor in the AC branch circuits feeding the equipment cabinets or racks.
- The TEBC must be uniform in size and not decrease in size as the grounding path approaches the earth.
- Metallic cable shields are not a substitute for a TEBC.
- There shall be a minimum separation of 2″ between a TEBC and power cables or telecommunication cables.
- There shall be a minimum separation of 2″ between a TEBC and ferrous materials where practical, or the TEBC must be bonded to the ferrous material.

Alternating Current Equipment Grounds (ACEGs)

An *alternating current equipment ground (ACEG)* is a conductor that bonds an electrical panel to a bonding busbar. If a telecommunication room or equipment room has an electrical panel that serves IT loads, the ACEG bonds the panel to the PBB or the SBB. The ACEG is bonded to the ACEG bus of the panel (if available) or the panel enclosure. Although not covered in ANSI/TIA-607-C, it is advisable to have a licensed electrician install an ACEG since it connects to the building electrical panel. Additional installation requirements include the following:

- An ACEG must be copper and may be insulated.
- An ACEG must be green or marked with a distinctive green color.
- The minimum size of an ACEG must be 6 AWG.
- An ACEG must be uniform in size and not decrease in size as the grounding path approaches the earth.
- An ACEG should not be installed in a ferrous metal conduit.
- If it is necessary to install an ACEG in a ferrous metal conduit, a bonding bushing must be used at each end to bond it to the conduit with a minimum 6 AWG conductor.

Telecommunications Equipment Bonding Conductors (TEBCs)

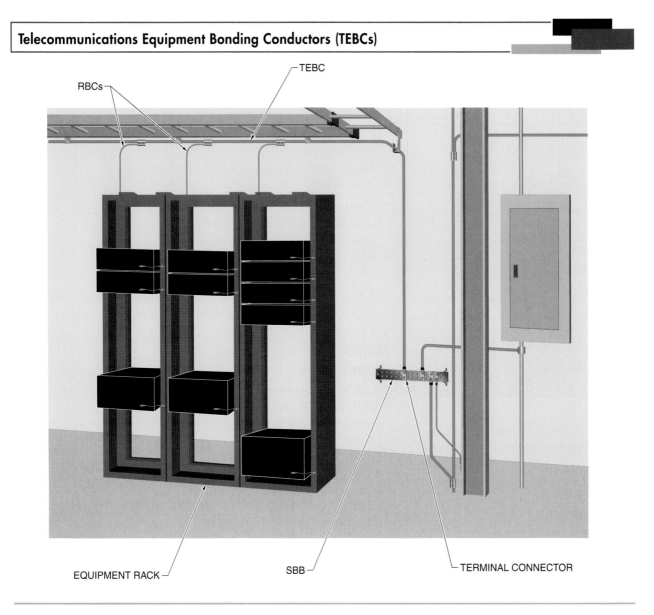

Figure 8-13. A telecommunications equipment bonding conductor (TEBC) bonds equipment racks to a PBB or SBB via rack bonding conductors (RBCs).

Summary

VDV systems are required to be grounded and bonded to protect personnel from harm and VDV system equipment from damage. Grounding and bonding requirements are written in the NEC®. The ANSI/TIA-607-C standard contains additional grounding methods that are required by telecommunication designers and engineers for many installations. VDV technicians must be familiar with the grounding requirements of the NEC® and ANSI/TIA-607-C.

Chapter Review

1. How do proper grounding and bonding protect personnel and limit damage to VDV systems?

2. What is bonding?

3. Which metallic parts of a VDV system are bonded?

4. An intersystem bonding termination is required by the NEC® to provide at least how many terminals for intersystem bonding conductors?

5. Which articles of the NEC® contain specific information on the grounding and bonding of VDV systems?

6. What is the minimum size required by the NEC® for a bonding conductor between a grounding electrode system for a building electrical service and a separate grounding electrode for a VDV system?

7. What are the three methods specified in ANSI/TIA-607-C for connections to a primary bonding busbar?

8. What are the minimum and maximum sizes of a telecommunication bonding backbone per ANSI/TIA-607-C?

9. What is the function of a backbone bonding conductor?

10. How are rack bonding conductors connected to telecommunication equipment bonding conductors?

11. What is an intersystem bonding termination?

12. What is transient voltage?

13. What is a primary protector?

Chapter Review

14. In what month and year did the EIA cease operation?

15. What is a primary bonding busbar (PBB)?

16. What is the recommended minimum size for a PBB insulator?

17. Why must a PBB have an insulator?

18. List ten components that must be connected to a PBB.

19. What is a telecommunications bonding conductor (TBC)?

20. What is a secondary bonding busbar (SBB)?

21. What is a telecommunications bonding backbone (TBB)?

22. What is a backbone bonding conductor (BBC)?

23. What is a telecommunications equipment bonding conductor (TEBC)?

24. What is an alternating current equipment ground (ACEG)?

25. Explain the difference between a grounding electrode (GE) and a grounding electrode conductor (GEC).

Chapter Activity Grounding and Bonding System Component Identification

ANSI/TIA-607-C, *Generic Telecommunications Bonding and Grounding (Earthing) for Customer Premises,* addresses the design and installation of a generic grounding and bonding system for telecommunications. *Identify the components using the information found on TIA-607-C in the chapter.*

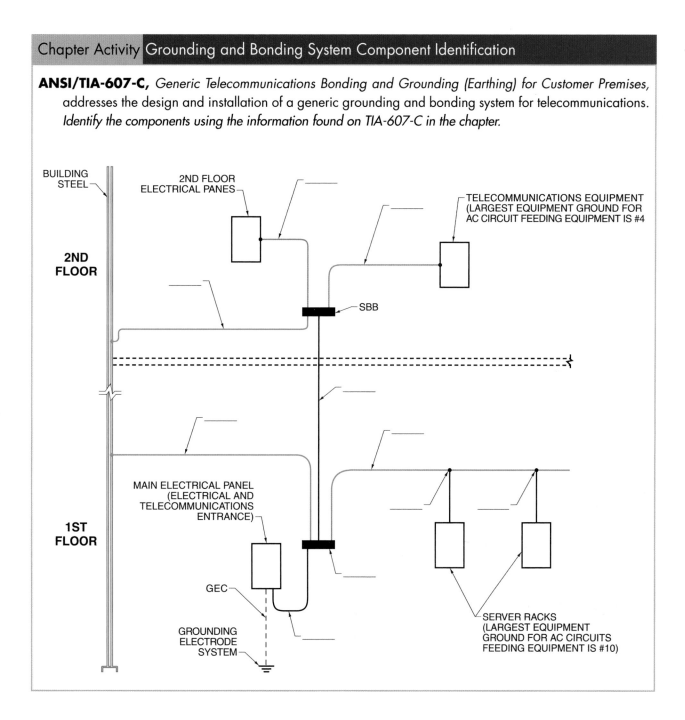

BUILDING STEEL

2ND FLOOR ELECTRICAL PANES

TELECOMMUNICATIONS EQUIPMENT (LARGEST EQUIPMENT GROUND FOR AC CIRCUIT FEEDING EQUIPMENT IS #4

2ND FLOOR

SBB

1ST FLOOR

MAIN ELECTRICAL PANEL (ELECTRICAL AND TELECOMMUNICATIONS ENTRANCE)

GEC

GROUNDING ELECTRODE SYSTEM

SERVER RACKS (LARGEST EQUIPMENT GROUND FOR AC CIRCUITS FEEDING EQUIPMENT IS #10)

Chapter Activity Grounding and Bonding System Component Identification

Using the list below, identify each component and conductor in the drawing. Some answers will be used more than once.

- Alternating Current Equipment Ground (ACEG)
- Telecommunications Bonding Conductor (TBC)
- Building Steel Bonding Conductor
- Backbone Bonding Conductor (BBC)
- Telecommunications Equipment Bonding Conductor (TEBC)
- Telecommunications Bonding Backbone (TBB)
- Secondary Bonding Busbar (SBB)
- Primary Bonding Busbar (PBB)
- Rack Bonding Conductor (RBC)

1. What is the minimum size of the TBB in the drawing?

2. What is the minimum size of the TBC in the drawing?

3. What is the minimum size of the TEBC for the first floor?

4. What is the minimum size of the TEBC for the second floor?

5. According to the TIA standard, what is the thickness and width of the PBB?

6. According to the TIA standard, the PBB and SBB should be made of what material?

7. What is the minimum size insulator used when mounting a PBB or SBB?

8. Is a BBC required for the building?

Chapter 9

Installation and Termination of Copper VDV Systems

VDV technicians spend the majority of their workday installing and terminating VDV systems with copper cables. These types of systems are used for phone, data, and video services in new installations and retrofits. Retrofits involve moves, additions, and changes (MAC) to existing installations.

The two primary types of copper cable used in VDV systems are four-pair twisted cable and coaxial cable, and there are different varieties of both cable types depending upon system requirements. The three steps involved with the installation and termination of copper VDV systems are rough-in, cable installation, and cable termination. Ideally, these steps would occur in order. However due to unforeseen job circumstances, such as back-ordered items or accelerated schedules, the order of these steps may change.

The installation and termination of copper VDV systems is governed by codes, standards, specifications, and the requirements of cable and related equipment OEMs. Improper installation and termination of a copper VDV system can result in inferior system performance. Furthermore, the system may not pass certification tests.

OBJECTIVES

- Describe how to prepare a job site for installation of VDV equipment per specifications.
- Describe how to prepare and pull cable as required per job site specifications.
- List the different firestopping methods.
- Explain the termination procedure for twisted-pair cables, connectors, and related devices.
- Explain the termination procedure for coaxial cables and connectors.

Digital Resources

ATPeResources.com/QuickLinks
Access Code 838502

CABLE INSTALLATION ROUGH-IN

Rough-in is the first step when installing and terminating a copper VDV system. The rough-in step involves preparing the telecommunications room or equipment room, the pathways, and the work areas for cable installation. **See Figure 9-1.** The telecommunications room or equipment room is the location from which the cables originate. The pathways serve as the route the cables take from the equipment room or telecommunications room to the work areas. The work areas are the locations where the cables terminate to telecommunications outlets.

When preparing a location (job site) for rough-in procedures, the project specifications and prints must be referenced to determine the proper placement of cables, receptacles, and related equipment. The specifications should include a design plan that can be used for this purpose.

Tech Tip

When roughing-in an area, accurate measurements for long distances can be attained with either an electronic distance measurement tool or a distance measuring wheel that automatically records a walked distance.

VDV Cable Installation

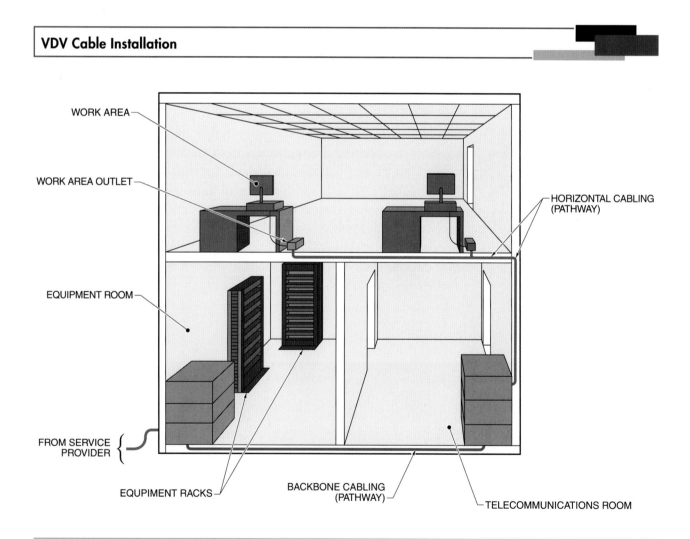

Figure 9-1. The rough-in step involves preparing the telecommunications room or equipment room, the pathways, and the work areas for cable installation.

Telecommunications Room and Equipment Room Rough-In

A *telecommunications room (TR)* is an enclosed space that houses telecommunications equipment, cable terminations, and cross-connect cabling used to service work areas on the same floor of a building. An *equipment room (ER)* is a central space that serves telecommunications equipment for occupants of a building or a campus of several buildings. **See Figure 9-2.** The terms telecommunications room and equipment room are often used interchangeably. However, there are differences between each type of room. A TR's primary function is to serve as a connection point between backbone and horizontal cabling. Cable terminations, cable cross-connects, and telecommunications equipment entrance cables are located in TRs. An ER's primary function is to serve as a location for telecommunications equipment. In addition, an ER may also include some or all of the functions of a TR. The rough-in for telecommunications and equipment rooms includes installing backboards, equipment racks, patch panels and cords, and cable-management systems. Key points concerning TRs include the following:

- A TR is required on the first floor of a multistory building.
- A smaller TR is required on each floor above the main floor in a multistory building.
- TRs are typically smaller than ERs.
- TRs are dedicated to the telecommunications functions of a building.

Telecommunications Rooms and Equipment Rooms

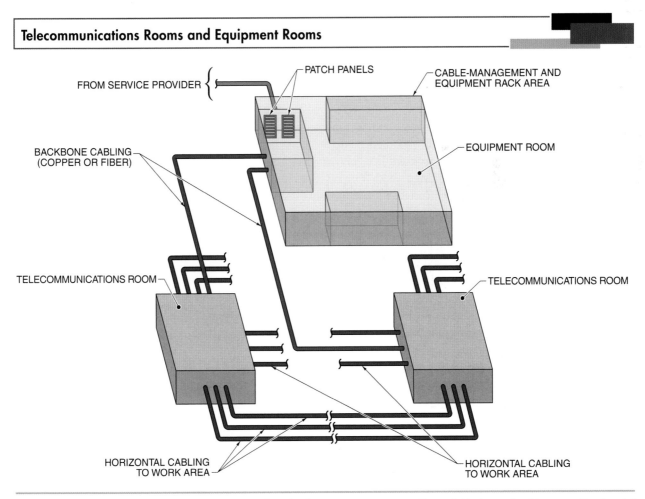

Figure 9-2. A telecommunications room's primary function is to serve as a connection point between backbone and horizontal cabling, while an equipment room's primary function is to serve as a location for telecommunications equipment.

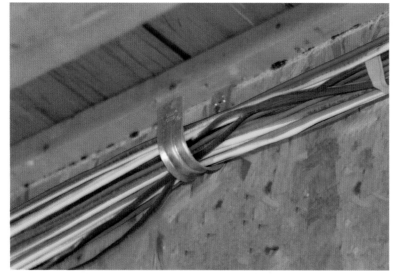

J-hooks are devices used to route VDV cables throughout a structure.

Typically, plywood sheets, or backboards, are installed in TRs and ERs. These backboards provide surfaces to permanently mount telecommunications equipment, cable-termination blocks, and cable-management components. Two walls of each room are covered with ¾" fire-rated plywood backboards. The International Building Code (IBC) refers to fire-rated plywood as "fire-retardant-treated wood." Fire-retardant-treated wood is treated with chemicals that slow the spread of flames in the event equipment mounted to it catches fire. If nonfire-rated plywood is used, it must be painted with fire-resistant paint.

As part of rough-in, equipment racks and telecommunications equipment such as cable-termination blocks and patch panels are installed. Depending on the installation requirements, equipment racks may be either mounted to the plywood backboard or freestanding and secured to the floor. Equipment room detail drawings included with VDV prints provide specific information.

Cable-management systems can be made up of many different products meant to neatly route cables to their proper locations. Cable-management systems are installed as part of telecommunications and equipment room rough-in. Devices such as D-rings,

J-hooks, and cable ducts are used to manage, organize, and route cables. Wire mesh cable trays or ladder trays are used to manage and route overhead cables entering telecommunications or equipment rooms. Cable racks are standard-size, metal, free-standing devices into which equipment is mounted. A 19" width across the front is the standard size for racks in the telecommunications industry. **See Figure 9-3.**

Cable-Management Systems

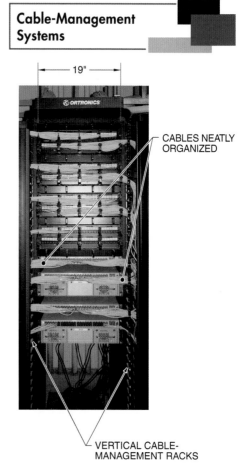

19"

CABLES NEATLY ORGANIZED

VERTICAL CABLE-MANAGEMENT RACKS

Figure 9-3. Cable-management systems are installed as part of telecommunications and equipment room rough-in.

Pathway Preparation

A *pathway* is a set of devices used to route telecommunications cables from a telecommunications room or equipment room to a work area. Pathways include conduit, cable tray, ducting, and plenum. **See Figure 9-4.**

A variety of raceways are used to route and support cables from a telecommunications room or equipment room to a work area. Raceways include J-hooks, metal conduit, plastic conduit, metal surface raceways, plastic surface raceways, ladder trays, wire mesh cable trays, and cellular metal floor raceways. The type of raceway used varies with the type of construction and the customer requirements. Often, more than one method is used. For example, wire mesh cable trays and J-hooks serve the same purpose.

Pathways

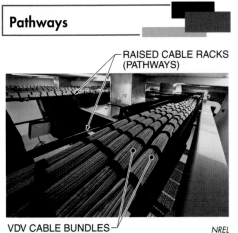

RAISED CABLE RACKS (PATHWAYS)

VDV CABLE BUNDLES   *NREL*

Figure 9-4. A pathway is a set of devices used to route telecommunications cables from a telecommunications room or equipment room to a work area.

A cable support system must be in place before a cable can be installed. Normally, VDV technicians install cable support systems. However in some projects, electricians may install the system on jobs that require extensive use of conduit. *Note:* The NEC® does not allow ceiling support wires or any part of a ceiling support system to be used to support VDV cables. Cables cannot be laid directly on or block ceiling tiles.

Work Area Preparation

A work area is a location where VDV cables terminate to telecommunications outlets that are used by building occupants. Endpoints must be established at work areas before

cables can be pulled. Mounting brackets, low-voltage rings, boxes, monuments for raised floors, and floor boxes are installed in work areas. The mounting devices may be attached to vertical studs, into holes cored to access cellular floors, or in boxes cut into tiles of a raised floor. Cable is pulled to these mounting devices. Telecommunication cable is then terminated to the outlets, and the outlets are attached to the mounting devices. Work areas in a building typically include open-office spaces, such as cubicles and individual offices. **See Figure 9-5.**

CABLE INSTALLATION

After cable rough-in is complete, cable installation can begin. The three steps to the cable installation process are cable preparation, cable pulling, and cable penetration firestopping. The steps must be performed in order. Cable installation typically consumes the greatest amount of time during a VDV installation. On jobs with complex or long cable runs, additional technicians may be required for a cable installation. After the three steps of cable installation are complete, the cable is ready for termination.

Cable Preparation

As part of cable preparation, a VDV technician must verify that the length of the entire run does not exceed the limits for four-pair twisted cable. Standards for four-pair twisted cable specify the maximum length of a permanent link as 90 m and the maximum length of a channel as 100 m. A *permanent link* is the installed cable, connectors, cross-connects, and outlets in a telecommunications installation project. A *channel* is the end-to-end transmission or communications path over which application-specific equipment is connected. A channel includes all permanent link elements and patch cords at each end. The standards allow 5 m of patch cord on each end. These lengths do not apply to coaxial cable. Typically, coaxial cable can be run longer distances than four-pair twisted cable.

Work Areas

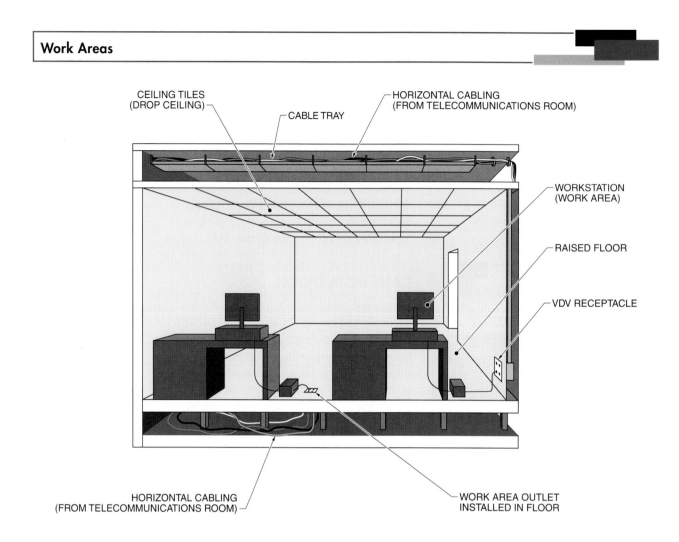

CUBICLES

INDIVIDUAL OFFICES

Figure 9-5. Work areas in a building typically include open-office spaces, such as cubicles and individual offices.

The next task a VDV technician must perform is to arrange the cables to be pulled. Cable is packaged by OEMs in self-dispensing boxes or reels. The advantages of boxes are that they are relatively lightweight, self-dispensing, and do not require a reel stand. The disadvantage of boxes is that they contain a shorter length of cable than reels. More cable runs are possible with a reel due to the greater amount of cable it contains. Boxes are normally used for small jobs, and reels are used for large jobs. Boxes or reels of cable are set up in a TR or ER when pulling cable.

After the cable is set up, the individual cables must be labeled. On some jobs, a permanent marker may be used to mark the cable sheath and its respective box or reel with the unique designation obtained from the VDV prints. *Note:* Handwritten labels are not acceptable as a final label for cable or telecommunication outlets. The opposite end of a cable is marked after the cable run is complete.

On large jobs, two sets of preprinted self-adhesive labels are generated by the VDV contractor. Each set has two identical labels, one for each end. One set of labels is used to mark both ends of the cable during the installation. The other set is used to mark both ends of the cable during the termination, as the cable is shortened and the initial label is lost with the discarded excess cable. Labels can also be generated by small handheld label makers in the field. **See Figure 9-6.**

In addition to the OEM information and cable type printed on the jacket, most cable is marked every other foot with its length. Because cable lengths in a box or reel can vary, a VDV technician must accurately track the starting length of cable for each box or reel supplied by the OEM. By tracking the cable length in each box or reel, the technician can track the amount of remaining cable in each to avoid being short on cable during a run.

Cable Identification

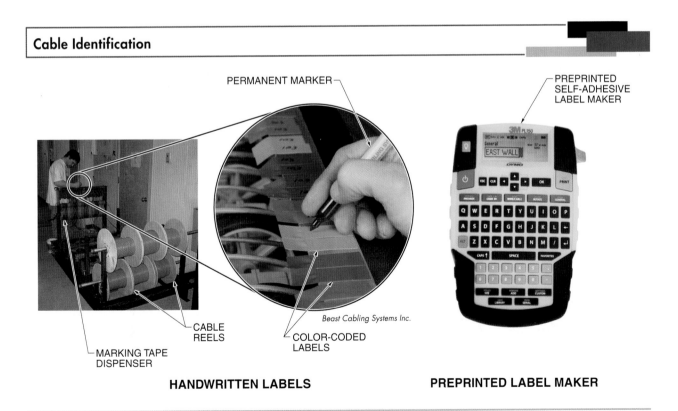

PERMANENT MARKER

PREPRINTED SELF-ADHESIVE LABEL MAKER

Beast Cabling Systems Inc.

MARKING TAPE DISPENSER

CABLE REELS

COLOR-CODED LABELS

HANDWRITTEN LABELS

PREPRINTED LABEL MAKER

Figure 9-6. On some jobs, a fine-point permanent marker may be used to mark the cable sheath and its respective box or reel with the unique designation obtained from the VDV prints.

Greenlee Textron, Inc.

Many OEMs have developed equipment that allows cable-pulling tasks to be performed with minimal damage to the cable and with less effort.

Cable Pulling

Often, a VDV technician must pull cable from a ceiling or open space through a raceway and a box near a VDV receptacle. There can be several different cable-pulling scenarios. The most common that are encountered by a VDV technician are the following:

- Cable may be pulled through J-hooks or loops in open air. The cable can be concealed above a ceiling or exposed below a ceiling. If installed above a ceiling in a commercial environment, the ceiling is typically accessible. *Note:* J-hooks and loops are devices specifically engineered for the purpose of routing VDV cables. **See Appendix.**

- Cable may be pulled through a ladder tray or wire mesh cable tray. The tray can be concealed above a ceiling or exposed below a ceiling. If installed above a ceiling in a commercial environment, the ceiling is typically accessible.

- Cable may be pulled through a raceway. The raceway can be metal conduit, plastic conduit, or a cellular floor metal raceway. It may be concealed or exposed.

Many installations use a combination of raceways to route cable from a TR to a work area. For example, a combination of wire mesh cable tray and J-hooks can be used, or a combination of wire mesh cable tray and conduit or VDV raceway. **See Figure 9-7.** Although each cable-pulling scenario is unique, there are common guidelines that apply to all. These guidelines for cable pulling fall into two categories, handling and separation. Handling guidelines include the following:

- When pulling cable, the maximum tension should not exceed 25 ft-lb. *Note:* The cable should not be jerked when performing pulling operations. Jerking the cable can exceed its tensile strength rating and cause permanent damage to it or to internal members.

- The cable must not be stepped on or run over with wheeled carts.

- The minimum bend radius for four-pair twisted cable should be four times the cable diameter. The minimum bend radius for twisted-pair cable with more than four pairs should be ten times the cable diameter. *Note:* It is a good practice to verify the minimum bend radius with the OEM.

- Cable ties must not be used to bundle or dress cables. Hook-and-loop (Velcro®) ties or straps must be used instead. Hook-and-loop ties and straps will not crush or deform the cable, which can result in system performance degradation. The ties and straps should be wrapped snug around the cable bundle but still allow the cables to slide back and forth. An additional advantage of hook-and-loop ties and straps is that they are reusable.

- Large heavy cables should always be placed at the bottom of a cable tray or bundle to prevent them from crushing smaller cables.

- When pulling cable, a pull line must always be used. Fold the cable over to form a loop and tape the cable to itself. Then attach a pull line to the loop.

Typically, a high-strength plastic rope or twine is used. Pull line is usually packaged in a 5 gal. plastic bucket with a length of several thousand feet. It is dispensed through a hole in the lid.

- Each end of the cable must have slack.

The second category of guidelines for cable pulling involves separation of cables from sources of EMI or physical damage. Separation guidelines include the following:

- VDV cables should not be run near sources of heat, such as light fixtures or pipes that carry hot liquids.
- VDV cables should not be run near sources of electromagnetic interference (EMI). *Electromagnetic interference (EMI) is interference in signal transmission or reception caused by the* radiation from electric and magnetic fields. Common sources of EMI include electric motors, transformers, and unshielded electric power conductors.

- A minimum 2″ space of separation must be maintained between VDV cables and unshielded electric power conductors, such as Romex® cables. If it is necessary to cross unshielded electric power conductors, they must cross at a 90° angle to minimize electrical interference.
- VDV cables must never be run in the same raceway with electric power conductors. In certain types of raceways however, such as surface metal raceways or power poles, VDV cables and electric power conductors may be run together provided there is a continuous metal barrier separating the two.

Cable Pulling

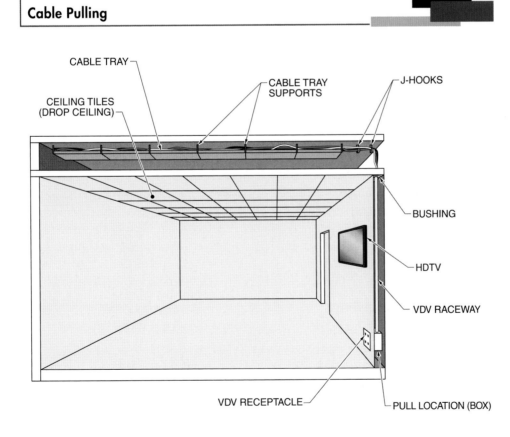

Figure 9-7. Often, a VDV technician must pull cable from a ceiling or open space through a raceway and a box near a VDV receptacle.

Open-Air Cable Pulling. Open-air cable pulling involves pulling cable through a series of J-hooks, rings, and loops rather than through enclosed cable trays. **See Figure 9-8.** This is a common installation for small-to-medium commercial projects. The following factors need to be considered when pulling cable through open air:

- The cable must be supported a minimum of every 4′ to 5′ on straight runs as well as at each turn. This is done with J-hooks installed on studs.
- The number of cables in each support should be limited to the maximum recommended by the OEM.
- If the cable is run above a suspended ceiling used as an air handling space, plenum-rated cable must be used.

- A pull line run through the supports must be used to pull the cable. A telescoping fiberglass pole (glow rod) can be used to aid the pulling process.
- Every turn is a pulling point, requiring a technician to handle the cable. Cable must not be pulled around turns without assistance as the cable may be damaged.

Cable Tray Cable Pulling. Cable tray cable pulling involves pulling cable through a solid bottom cable tray, a wire mesh cable tray, or a ladder tray. **See Figure 9-9.** A cable tray can have bends and offsets along its route, and power poles are often used to route cable from the tray to a work area. This is a common installation for medium-to-large commercial projects. The following factors need to be considered when pulling cable through a cable tray:

Open-Air Cable Pulling

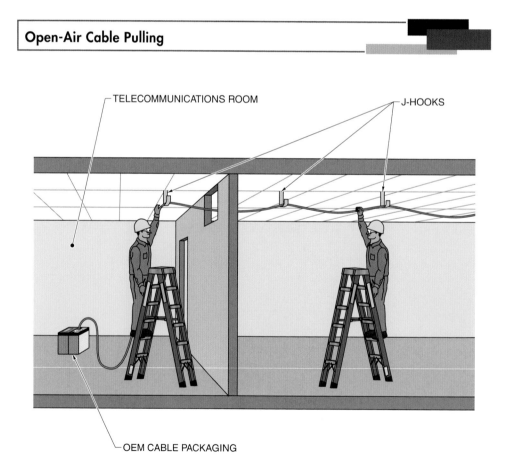

TELECOMMUNICATIONS ROOM

J-HOOKS

OEM CABLE PACKAGING

Figure 9-8. Open-air cable pulling involves pulling cable through a series of J-hooks, rings, and loops.

- Cable tray can be installed above a suspended ceiling, exposed, or under a raised computer floor.

- If the cable tray is run above a suspended ceiling or under a raised computer floor that is used as an air handling space, plenum-rated cable must be used in the tray.

- The TIA recommends limiting the fill of a tray to 50%. The cable tray OEM or VDV detail drawings provide a chart with the maximum fill for various cable types. The number of cables must be limited to the maximum recommended by the tray OEM.

- A pull line that is run through a tray must be used to pull the cable.

- Every turn is a pulling point and requires the use of a pulley or a VDV technician to aid with the pull. Attempting to pull around turns without assistance may damage the cable.

Cable Tray Cable Pulling

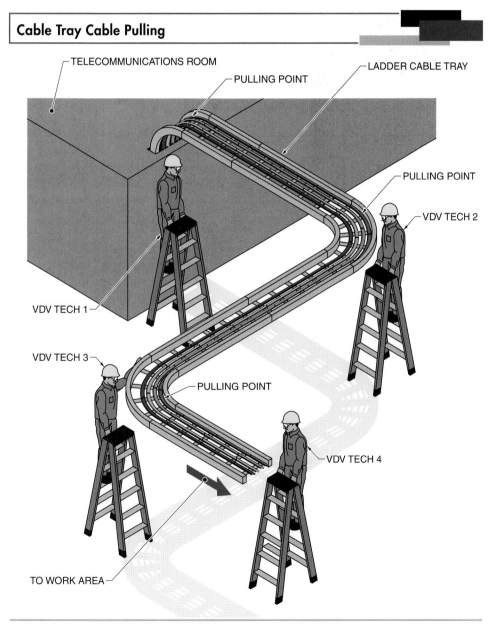

Figure 9-9. Cable tray cable pulling involves pulling cable through a solid bottom cable tray, a wire mesh cable tray, or a ladder tray. VDV technicians or pulleys are placed at each pulling point to aid in the pull.

Raceway Cable Pulling. Raceway cable pulling involves pulling cable through conduit. **See Figure 9-10.** There are several different types of conduit that can be used including, but not limited to, rigid metal conduit (RMC), intermediate metal conduit (IMC), electrical metallic tubing (EMT), flexible metal conduit (FMC), and rigid polyvinyl chloride conduit (PVC). Conduit can have bends, offsets, and junction boxes. This is a common installation for large commercial projects and industrial projects. The TIA standards contain the following factors that need to be considered when pulling cable through conduit:

- The smallest conduit that can be installed on commercial or industrial projects is ¾″ trade size.
- Per the TIA, only two 90° bends are allowed in a conduit run. If the conduit run requires more than two, a junction box must be installed.

- Per the TIA, the maximum length allowed for a conduit run is 100′. If the conduit run must be longer, a junction box must be installed.
- The recommended fill for conduit is 40%. Cable OEM literature or VDV detail drawings provide a chart with the maximum number for various cable types. The number of cables must be limited to the maximum recommended by the OEM.
- A fish tape can be used to install a pull line in the conduit. The pull line must be used to pull the cable.
- Each end of the conduit should have a push-on bushing or a connector with a bushing installed to protect the cable from abrasion.
- If it is necessary to use pulling lubricant (such as soap), only lubricant specifically designed for VDV cables must be used. General-purpose pulling lubricants may adversely affect cable performance.

Raceway Cable Pulling

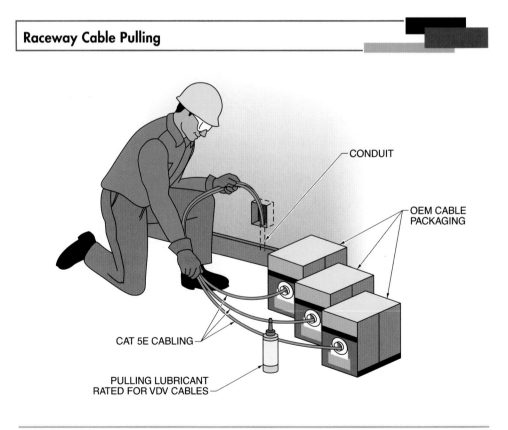

CONDUIT

OEM CABLE PACKAGING

CAT 5E CABLING

PULLING LUBRICANT RATED FOR VDV CABLES

Figure 9-10. Raceway cable pulling involves pulling cable through conduit.

Cable Penetration Firestopping

Structures are constructed to comply with various building codes. These building codes include provisions for fire-resistant or fire-rated ceilings, floors, and walls. The reason for fire-resistant or fire-rated construction is to prevent the spread of fire, or smoke and harmful gases caused by a fire, from one area of a structure to another. Containing fire, smoke, and gases provides additional time for occupants to safely exit a building and firefighters to extinguish the fire. Whenever a fire-rated floor, wall, or other element is penetrated, the penetration must be properly sealed in order to maintain the fire rating.

One of the most important elements of a VDV technician's job is firestopping. *Firestopping* is the process of sealing any penetration in a fire-rated building element in order to maintain its rating.

The two general categories of penetrations are through penetrations and membrane penetrations. A *through penetration* is a penetration that passes completely through a fire-rated element. For example, a VDV cable that passes through a fire-rated floor is a through penetration. A *membrane penetration* is a penetration that only penetrates one side of a fire-rated element. For example, an electrical outlet box on one side of a fire-rated wall is a membrane penetration.

Firestopping material is used to seal any gaps or holes. The material is engineered to return the ceiling, floor, or wall to its original fire rating. A wide variety of firestopping materials are available. Firestop materials include blocks, caulks, foams, mortars, putties, putty pads, putty sticks, and sprays. A label documenting the firestopping installation is required adjacent to each firestop location. The label typically includes information such as the firestop product trade name, the installation date, the rating, and the name and contact information for the installer. **See Figure 9-11.**

Firestopping material must be tested and approved by a recognized testing laboratory.

It must be installed per the manufacturer's instructions. The testing lab approval is void if the instructions are not followed. On large jobs, the VDV prints have detail drawings describing the materials and methods for firestopping in a variety of situations, such as concrete floor penetrations, concrete wall penetrations, and drywall penetrations. On smaller jobs, the VDV technician must determine the appropriate firestopping methods and materials.

CABLE TERMINATION

Copper VDV cables (twisted pair and coaxial) originate in a TR and are routed through various raceways and supports to work areas. A copper VDV cable must be terminated at both ends, at the TR and the work area. In the TR, cables are dressed and routed to their designated jacks, patch panels, and cross-connects. The cables are then labeled and terminated. This process is repeated in each work area. The majority of copper VDV installation issues involve termination errors. A VDV technician must take precautions and be detailed when terminating cables in order to avoid errors.

Beast Cabling Systems, Inc.

Cables are typically organized and labeled prior to being routed to the work areas in a building.

Cable Penetration Firestopping

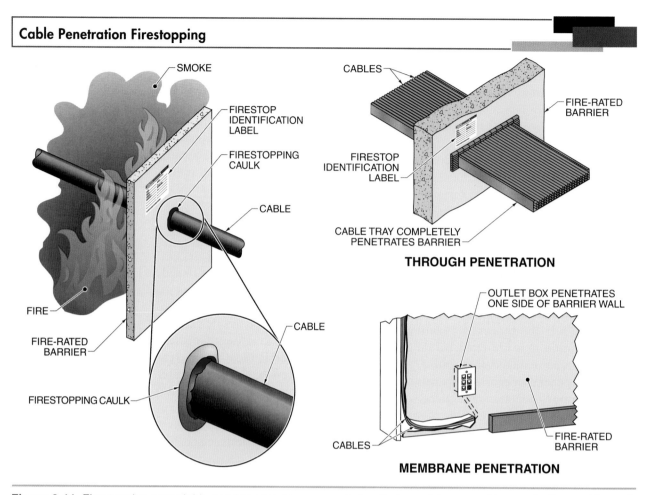

Figure 9-11. Firestopping material is used to seal any gaps or holes that result from a cable penetration and is engineered to return a penetrated ceiling, floor, or wall to its original fire rating.

Twisted-Pair Cable Termination

The most common termination performed by a VDV technician involves twisted-pair cable, specifically four-pair twisted cable. VDV technicians need to understand twisted-pair color codes, cable categories, hardware categories, and conductor termination patterns. In addition, a VDV technician must properly manage and label twisted-pair cable.

Each conductor pair has a tip conductor and a ring conductor. A *tip conductor* is the first wire in a pair of wires. A *ring conductor* is the second wire in a pair of wires. Tip and ring conductor pair colors complement each other for easy identification. For example, the first conductor of the first pair (tip conductor) is white with a blue stripe, and the second conductor (ring conductor) is blue with a white stripe. Tip conductors may be solid white or striped. The terms "tip" and "ring" conductor

come from the early telephone system switchboards. The wire at the end of the plug was called the "tip," and wire at the connecting point of the jack was called the "ring."

Twisted-Pair Color Codes. The color code for four-pair twisted cables is based on a system of 25-pair (50-conductor) telephone cables using colored conductors with stripes to designate specific pairs of conductors. The tip colors are white for pairs 1 to 5, red for pairs 6 to 10, black for pairs 11 to 15, yellow for pairs 16 to 20, and violet for pairs 21 to 25. The ring colors are blue for the first pair in a group of five pairs, orange for the second pair in a group of five, green for the third pair in a group of five, brown for the fourth pair in a group of five, and slate for the fifth pair in a group of five. The tip conductors may have a stripe of ring conductor color to identify the specific pair.

The ring conductors may have a stripe of tip conductor color to identify the specific pair. *Note:* Some higher categories of cable, such as Cat 6 and Cat 6A, may omit stripes on the conductors because the tight twisting of the individual pairs allows identification. **See Figure 9-12.** Cables with more than 25 pairs use a system of colored binders to identify each group of 25. A *binder* is colored plastic tape or colored string-like fiber that is wrapped helically around a specific group of 25 twisted pairs. The color of the binder indicates the group. Only the color of the binders change, not the colors of the groups.

Twisted-Pair Color Codes

4-Pair Conductor Color Code

Pair		Jacket Color
1	Tip	White
	Ring	Blue
2	Tip	White
	Ring	Orange
3	Tip	White
	Ring	Green
4	Tip	White
	Ring	Brown

25-Pair Conductor Color Code

Color Group	Pair		Conductor Jacket Color	Stripe Color
White (1)	1	Tip	White	Green
		Ring	Blue	White
	2	Tip	White	Orange
		Ring	Orange	White
	3	Tip	White	Blue
		Ring	Green	White
	4	Tip	White	Brown
		Ring	Brown	White
	5	Tip	White	Slate
		Ring	Slate	White
Red (2)	6	Tip	Red	Blue
		Ring	Blue	Red
	7	Tip	Red	Orange
		Ring	Orange	Red
	8	Tip	Red	Green
		Ring	Green	Red
	9	Tip	Red	Brown
		Ring	Brown	Red
	10	Tip	Red	Slate
		Ring	Slate	Red
Black (3)	11	Tip	Black	Blue
		Ring	Blue	Black
	12	Tip	Black	Orange
		Ring	Orange	Black
	13	Tip	Black	Green
		Ring	Green	Black
	14	Tip	Black	Brown
		Ring	Brown	Black
	15	Tip	Black	Slate
		Ring	Slate	Black
Yellow (4)	16	Tip	Yellow	Blue
		Ring	Blue	Yellow
	17	Tip	Yellow	Orange
		Ring	Orange	Yellow
	18	Tip	Yellow	Green
		Ring	Green	Yellow
	19	Tip	Yellow	Brown
		Ring	Brown	Yellow
	20	Tip	Yellow	Slate
		Ring	Slate	Yellow
Violet (5)	21	Tip	Violet	Blue
		Ring	Blue	Violet
	22	Tip	Violet	Orange
		Ring	Orange	Violet
	23	Tip	Violet	Green
		Ring	Green	Violet
	24	Tip	Violet	Brown
		Ring	Brown	Violet
	25	Tip	Violet	Slate
		Ring	Slate	Violet

Color Code for Binder Groups

Pairs	Color
001–025	White/Blue
026–050	White/Orange
051–075	White/Green
076–100	White/Brown
101–125	White/Slate
126–150	Red/Blue
151–175	Red/Orange
176–200	Red/Green
201–225	Red/Brown
226–250	Red/Slate
251–275	Black/Blue
276–300	Black/Orange
301–325	Black/Green
326–350	Black/Brown
351–375	Black/Slate
376–400	Yellow/Blue
401–425	Yellow/Orange
426–450	Yellow/Green
451–475	Yellow/Brown
476–500	Yellow/Slate
501–525	Violet/Blue
526–550	Violet/Orange
551–575	Violet/Green
576–600	Violet/Brown

TIP CONDUCTOR — RING CONDUCTOR

TWISTED PAIR

Figure 9-12. The color code for four-pair twisted cables is based on a system for 25-pair (50-conductor) telephone cables using colored conductors with stripes to designate specific pairs of conductors.

Twisted-Pair Cable Categories. Twisted-pair cable is classified using a system of categories. The categories define the transmission rate the cable can support. Common categories include Cat 5e and Cat 6A. The components used to terminate twisted-pair cable and connect devices to the cabling infrastructure also have a category rating. These include wiring blocks, connecting blocks, patch panels, jacks, and patch cords. The category rating of all components and the cable must match. A mismatch between a cable and a termination component will result in substandard performance of the VDV cabling infrastructure. **See Figure 9-13.**

Termination Patterns. The TIA has specific conductor termination patterns when terminating twisted-pair cable to RJ-45 jacks, RJ-45 modular plugs, and patch panels. The two predominant conductor termination patterns for RJ-45 connectors are T568A and T568B (also known as AT&T 258A). **See Figure 9-14.** The T568A or T568B pattern is graphically shown on most termination components. The T568B pattern is more common in commercial installations. The T568A pattern is more common in residential installations. The VDV technician must always verify which termination pattern is required for each specific job. This information can be found on VDV prints and specifications or obtained by contacting the customer's telecom manager or IT manager.

Twisted-Pair Cabling Equipment

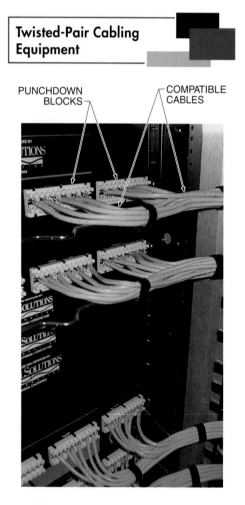

PUNCHDOWN BLOCKS

COMPATIBLE CABLES

Figure 9-13. The category rating of all twisted-pair cabling components in a VDV system must match the category rating of the cable.

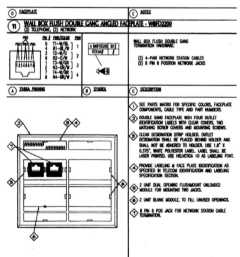

Twisted-pair termination patterns can be determined from referencing the prints and specifications for a specific installation.

Twisted-Pair Cable Management and Labeling. Cable management and labeling are important when terminating twisted-pair cable. Cable management involves dressing, routing, and supporting the cable in a neat and workmanlike manner. Cable labeling involves labeling the cable, work-area outlets, patch panels, cross-connects, and other termination components. **See Figure 9-15.**

T568A and T568B Terminations

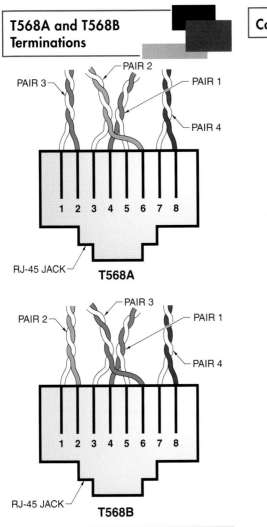

Figure 9-14. The two predominant conductor termination patterns for RJ-45 connectors are T568A and T568B.

Cable management at a work area outlet consists of leaving a sufficient amount of slack in the event retermination is necessary and ensuring the cable is not subject to physical damage. Generally, 12″ to 18″ of slack is considered adequate. The minimum bend radius of the cable must also be observed. Hook-and-loop straps should be used to secure the cable as needed. Care should be exercised not to overtighten the hook-and-loop fasteners and deform the cable. Deformed cable can result in performance degradation. Cable ties are not recommended as they can cause cable deformation over time.

Cable Management and Labeling

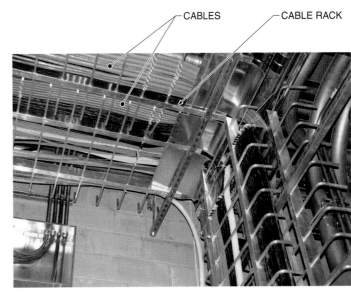

CABLE MANAGEMENT

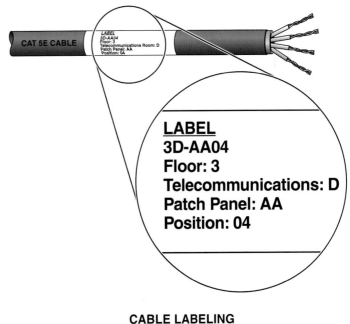

LABEL
3D-AA04
Floor: 3
Telecommunications: D
Patch Panel: AA
Position: 04

CABLE LABELING

Figure 9-15. Cable management involves dressing, routing, and supporting the cable in a neat and workmanlike manner, while cable labeling involves labeling the cable, work-area outlets, patch panels, cross-connects, and other termination components.

Cable management in a TR is similar to cable management at a work area. Cable management in a TR involves the following:

- A reasonable amount of slack should be left in the event retermination is necessary.
- The cable must be adequately supported and protected from physical damage, such as from sharp edges.
- When loops of extra cable are required, the loops should not be symmetrical, as symmetrical loops can degrade cable performance and lead to problems from excessive crosstalk.
- Hook-and-loop straps should be used to secure loops and cable bundles. The fasteners should be spaced randomly and not overtightened in order to prevent cable performance issues. Cable ties are not recommended for securing cables.
- Cable-management devices, both horizontal and vertical, must be used to neatly route the cable. Devices include vertical support bars, horizontal support bars, rings, saddles, and duct.
- The minimum bend radius of the cable must be observed.

The ANSI/TIA 606 standard covers the administration of a telecommunications infrastructure. Suggested labeling schemes are included as part of the standard. The customer or end user makes the final decision on the labeling scheme.

Normally, labels are applied before cables are terminated. Important points related to any labeling scheme include the following:

- Both ends of a cable, at a work area and in a TR, must be labeled within 12″ of the terminations.
- Labels must be machine generated. Handwritten labels are not acceptable.
- The faces of work area outlets must be labeled.
- The patch panels, cross-connects, and connecting blocks located in the TR must be labeled.

Twisted-Pair Cable Termination Devices

There are a number of different types of devices for terminating twisted-pair cable. These devices include 66 blocks, 110 wiring blocks and connecting blocks, 110-style patch panels, 110 RJ-45 jacks, and RJ-45-style modular patch panels. All of these devices rely on some type of insulation displacement contact (IDC) to terminate the conductors. With IDCs, a tool is used to force an insulated conductor of a twisted-pair cable into a groove between two metal contacts. The insulation is squeezed until the soft material is pushed to the side, and the conductor makes a connection with the metal contacts. In addition to seating the conductor, the tool cuts off any excess on the end. IDC contacts provide a consistent, efficient, and reliable termination method for twisted-pair cables. Legacy voice installations and some legacy data installations still use 66 blocks. The majority of new installations use 110 IDC termination devices for voice and data. **See Figure 9-16.**

When cables are installed, they must be adequately supported and protected from conditions that may cause physical damage such as sharp edges.

Tech Tip

Many OEMs provide modern cable-management systems for residential VDV installations.

66 and 110 Blocks

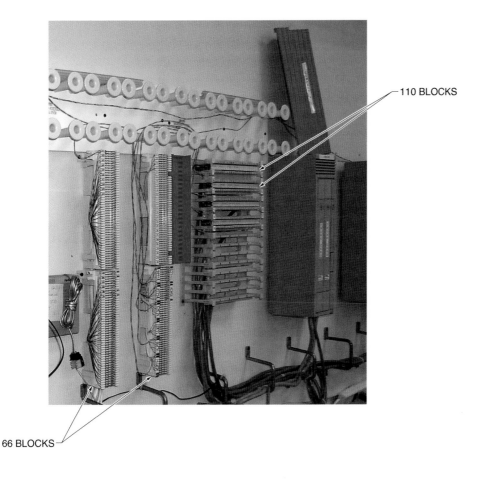

110 BLOCKS

66 BLOCKS

Figure 9-16. Legacy voice installations and some legacy data installations still use 66 blocks, while the majority of new installations use 110 IDC termination devices for voice and data.

The most common tool used for IDC terminations is an impact, or punchdown, tool. The tool typically has a removable, two-sided combination blade. One side of the blade is for 66-type IDC terminations, and the other side is for 110-type IDC terminations. The impact tool has an adjustment for impact, either "HI" or "LO." Always consult OEM literature or instructions for the recommended impact setting. The tool also has a compartment to store spare blades. Some impact tools have an integral wire spudger. A *wire spudger* is a device used to remove individual wires from a bundle, push wires into position, or remove wires from an IDC. Wire spudgers are sometimes referred to as wire picks. Other tools used are round cable cutters, cable strippers, and electrician's scissors. **See Figure 9-17.**

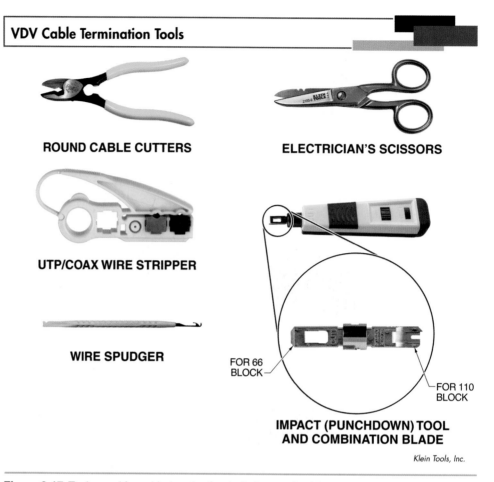

VDV Cable Termination Tools

ROUND CABLE CUTTERS

ELECTRICIAN'S SCISSORS

UTP/COAX WIRE STRIPPER

WIRE SPUDGER

FOR 66 BLOCK

FOR 110 BLOCK

IMPACT (PUNCHDOWN) TOOL AND COMBINATION BLADE

Klein Tools, Inc.

Figure 9-17. Tools used for cable termination include round cable cutters, electrician's scissors, UTP/coax wire strippers, impact tools and blades, and wire spudgers.

66 Block Termination. For many years, 66 blocks were standard technology. These blocks were originally designed to splice two 25-pair cables. Normally, a 66 block snaps into a raised plastic mounting unit referred to as a standoff. The standoff is then mounted to a backboard and provides a channel to route wires. The most common 66 block has four columns of 50 insulation displacement contacts when viewed vertically. The two columns on the left are designated for one 25-pair cable, and the two columns on the right are designated for another 25-pair cable. The two insulation displacement contacts in each row in a column are internally connected. There is a fanning strip or wire guide for each wire on either side of the 66 block. **See Figure 9-18.**

Terminating a 25-pair cable to a 66 block requires a VDV technician to remove the minimum amount of cable jacket necessary to fan the wires out to their designated IDC terminations. The twist on the pairs must be maintained up to the fanning strip. Then the individual wires must be routed through their respective slots to the closest IDC terminations. The VDV technician then positions the wires for termination. The tip conductors terminate on the odd-numbered rows, and the ring conductors terminate on the even-numbered rows. Typically, the excess wire that faces down is cut off.

After all of the wires are in position for termination, an impact tool is used to terminate the wires. The "CUT" side of the 66 blade must be positioned to cut off excess wire. If the "CUT" side is positioned incorrectly, the conductor is cut from the wrong side instead of being terminated.

66 Blocks

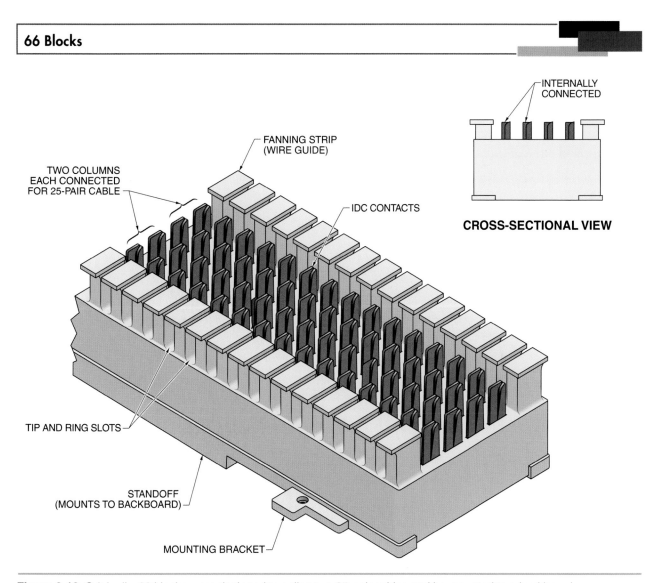

INTERNALLY
CONNECTED

CROSS-SECTIONAL VIEW

FANNING STRIP
(WIRE GUIDE)

TWO COLUMNS
EACH CONNECTED
FOR 25-PAIR CABLE

IDC CONTACTS

TIP AND RING SLOTS

STANDOFF
(MOUNTS TO BACKBOARD)

MOUNTING BRACKET

Figure 9-18. Originally, 66 blocks were designed to splice two 25-pair cables and be mounted to a backboard.

Other conductors can be connected to a 25-pair cable using the adjacent, internally connected IDC termination, or the 25-pair cable can be connected to another 25-pair cable terminated on the opposite side using bridging clips. A *bridging clip* is a small piece of steel that snaps over the two center IDCs and thus connects (bridges) the two outer IDCs. **See Figure 9-19.**

Terminating four-pair twisted cable to a 66 block requires a VDV technician to strip the minimum amount of jacket from the cable while maintaining the jacket as close as possible to the termination. **See Figure 9-20.** The twist on the pairs should be maintained through the fanning strip. The first pair goes through the first slot of the strip, and the second slot of the fanning strip is skipped. Then the second pair goes through the third slot of the fanning strip, and the sequence is continued for the third and fourth pair. Thus, each pair is between their respective tip and ring IDC terminations, and the twist is maintained up to the termination. The tip conductors terminate on the odd-numbered rows, and the ring conductors terminate on the even-numbered rows.

Bridging Clips

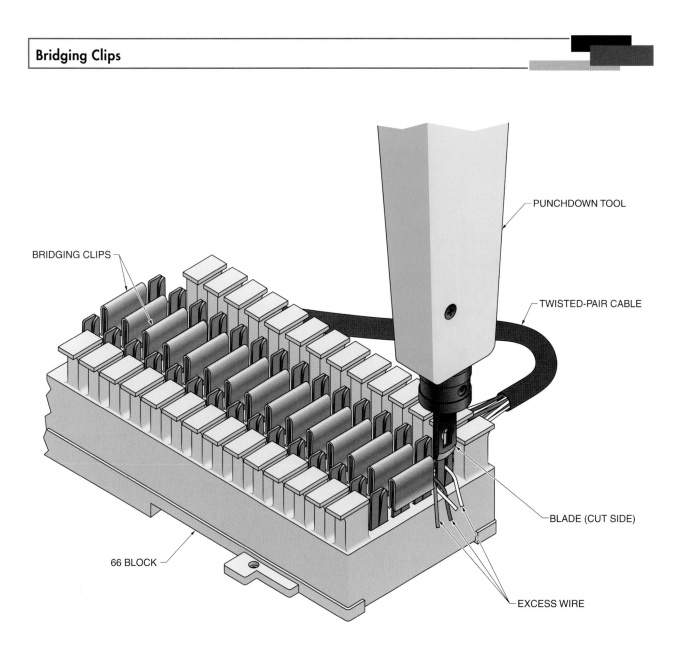

BRIDGING CLIPS

PUNCHDOWN TOOL

TWISTED-PAIR CABLE

66 BLOCK

BLADE (CUT SIDE)

EXCESS WIRE

Figure 9-19. A bridging clip is a small piece of steel that snaps over two 25-pair cables and can be mounted to a backboard.

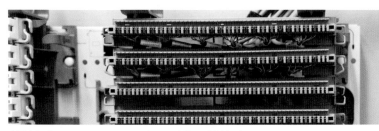

110 blocks are more commonly used than 66 blocks.

After all of the wires are in position for termination, an impact tool is used to terminate the wires. A 66 block can accommodate up to 12 four-pair cables (six cables on each side), leaving the bottom two rows unused. Other conductors can be connected to four-pair cables using the adjacent, internally connected IDC termination.

Terminating Four-Pair Twisted Cable to 66 Blocks

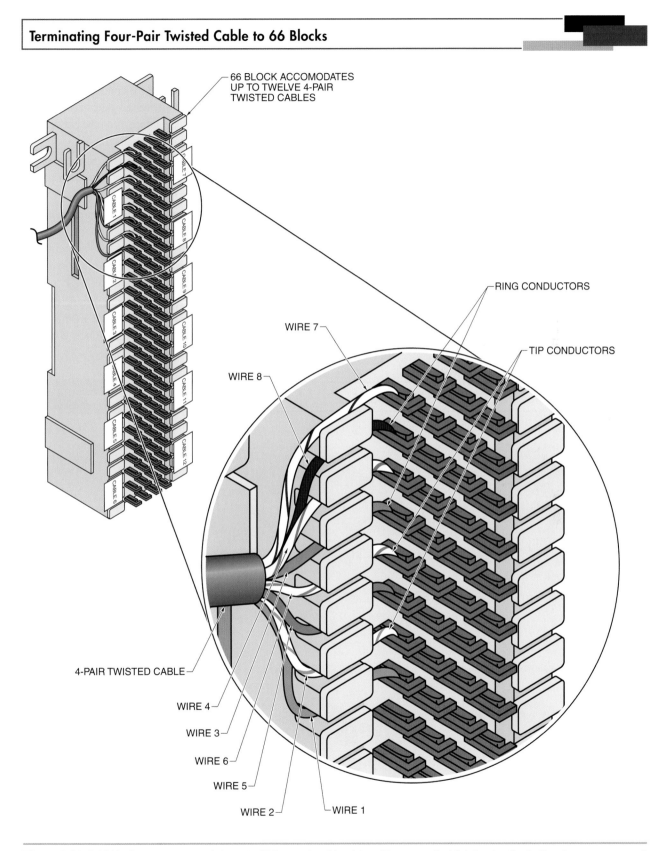

66 BLOCK ACCOMODATES
UP TO TWELVE 4-PAIR
TWISTED CABLES

RING CONDUCTORS

TIP CONDUCTORS

WIRE 7

WIRE 8

CABLE 1
CABLE 2
CABLE 3
CABLE 4
CABLE 5
CABLE 6
CABLE 7
CABLE 8
CABLE 9
CABLE 10
CABLE 11
CABLE 12

4-PAIR TWISTED CABLE

WIRE 4

WIRE 3

WIRE 6

WIRE 5

WIRE 2

WIRE 1

Figure 9-20. A 66 block can accommodate up to 12 four-pair cables (six cables on each side), leaving the bottom two rows unused.

110 Block Termination. A 110 wiring block and connecting block can be used to connect two wires. The 110 wiring block serves as a base and has multiple rows of 50 positions to accept 25 pairs. It typically has legs for support and room behind the block to route cables. The connecting block or clip has IDC terminations on both the bottom and top that are mechanically joined. The connecting block serves as the connecting mechanism between the two wires. Connecting blocks are available in two- to five-pair C clips for use in different applications. The following procedure can be used to terminate four-pair twisted cable using 110 wiring blocks and C-4 connecting blocks:

1. Route the cable to the row where it terminates.

2. Strip the minimum amount of jacket from the cable while maintaining the jacket as close as possible to the termination.

3. Fan the conductors out and lace them into their positions for proper termination. Generally, a maximum ½″ of untwist is permitted. Some OEMs only permit ¼″ of untwist. Be sure the conductors terminating on the top row have their excess fanned up, and the conductors on the bottom row have their excess fanned down.

4. Use an impact tool to terminate the conductors. Be sure the cut blade is in the proper direction.

5. Lace the second twisted-pair cable into the connecting block according to the color code following the same untwist guidelines, and set it in place with the impact tool. **See Figure 9-21.**

6. Align the connecting block or clip over the wiring block according to the color code and set in place with the impact tool.

Terminating Twisted Pair Cable to 110 Blocks

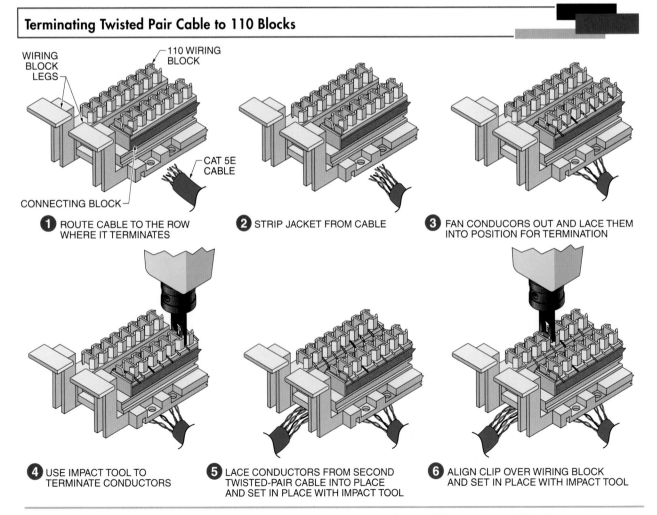

① ROUTE CABLE TO THE ROW WHERE IT TERMINATES

② STRIP JACKET FROM CABLE

③ FAN CONDUCORS OUT AND LACE THEM INTO POSITION FOR TERMINATION

④ USE IMPACT TOOL TO TERMINATE CONDUCTORS

⑤ LACE CONDUCTORS FROM SECOND TWISTED-PAIR CABLE INTO PLACE AND SET IN PLACE WITH IMPACT TOOL

⑥ ALIGN CLIP OVER WIRING BLOCK AND SET IN PLACE WITH IMPACT TOOL

Figure 9-21. A 110 wiring block serves as a base and has multiple rows of 50 positions to accept 25 pairs.

110 Patch Panel Termination. Many patch panels have RJ-45 jacks on the front that are factory connected to the back of 110-type connecting blocks on the rear. The other side of the connecting block is then used to terminate either horizontal or utility wiring. Short patch cords with RJ-45 connecters are then used to make connections between the jacks and the wiring. The following procedure can be used to terminate a 110-style patch panel:

1. Route the cable to the appropriate 110 block.
2. Use a data cable stripper to strip the minimum amount of jacket from the cable while maintaining the jacket as close as possible to the termination.
3. Fan out pairs, and straighten each wire.
4. Lace the wires into the 110 block according to the color code. Generally, a maximum ½″ of untwist is permitted. Some OEMs only permit ¼″ of untwist. Be sure the conductors terminating on the top row have their excess fanned up, and the conductors on the bottom row have their excess fanned down.
5. Use an impact (punchdown) tool to terminate the wires into 110 block. **See Figure 9-22.**

110 RJ-45 Jack Termination. Work area devices such as PCs and printers are connected to a network with a patch or workstation cable inserted into an RJ-45 jack. Most work area outlets consist of 110 RJ-45 jack terminations. The jacks are terminated and then snapped into protective plates or covers.

These jacks are also used with patch panels. They are terminated and then snapped into a patch panel faceplate with multiple RJ-45 jack locations. Most patch panels are designed to be mounted within a standard 19″ wide rack.

110-Style Patch Panel Termination

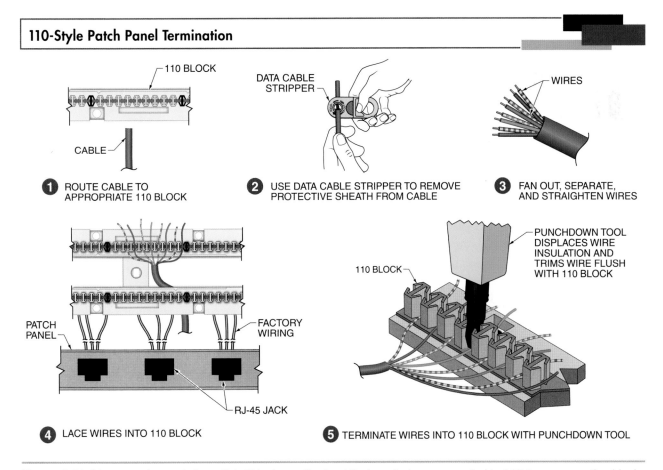

1 ROUTE CABLE TO APPROPRIATE 110 BLOCK

2 USE DATA CABLE STRIPPER TO REMOVE PROTECTIVE SHEATH FROM CABLE

3 FAN OUT, SEPARATE, AND STRAIGHTEN WIRES

4 LACE WIRES INTO 110 BLOCK

5 TERMINATE WIRES INTO 110 BLOCK WITH PUNCHDOWN TOOL

Figure 9-22. Many patch panels have RJ-45 jacks on the front that are factory-connected to 110-type connecting blocks on the rear.

The two termination tools for 110-type RJ-45 jacks are manufacturer specific and jack-design specific. A VDV technician must use the correct tool for each OEM's RJ-45 jack. The first type is an impact tool, and the second type is a termination tool designed to terminate all eight conductors simultaneously and thus save time. Typically, the tool is designed to work with a specific OEM's RJ-45 jack. The following procedure can be used to terminate a 110-type RJ-45 jack using either method:

1. Use a data cable stripper to strip the minimum amount of jacket from the cable while maintaining the jacket as close as possible to the termination.

2. Fan out, separate, and straighten the wires. Lace the straightened wires into the RJ-45 jack according to the color code. Generally, a maximum ½″ of untwist is permitted. Some OEMs only permit ¼″ of untwist.

3. Route the wires into the connector.

4. Terminate the wires into the connector with an impact (punchdown) tool or jack tool.

5. Snap the RJ-45 jack into the wall plate or patch panel faceplate.

Use either an impact (punchdown) tool to terminate the conductors individually or a jack tool (a hand tool designed for terminating VDV jacks) to terminate all the conductors simultaneously. Some jack tools can be pistol shaped. **See Figure 9-23.**

110 RJ-45 Jack Termination

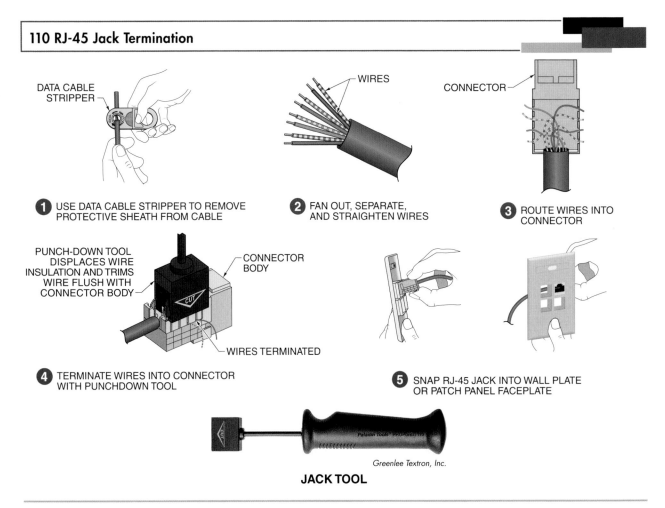

DATA CABLE STRIPPER

WIRES

CONNECTOR

1 USE DATA CABLE STRIPPER TO REMOVE PROTECTIVE SHEATH FROM CABLE

2 FAN OUT, SEPARATE, AND STRAIGHTEN WIRES

3 ROUTE WIRES INTO CONNECTOR

PUNCH-DOWN TOOL DISPLACES WIRE INSULATION AND TRIMS WIRE FLUSH WITH CONNECTOR BODY

CONNECTOR BODY

WIRES TERMINATED

4 TERMINATE WIRES INTO CONNECTOR WITH PUNCHDOWN TOOL

5 SNAP RJ-45 JACK INTO WALL PLATE OR PATCH PANEL FACEPLATE

Greenlee Textron, Inc.

JACK TOOL

Figure 9-23. Most work area outlets consist of 110-type RJ-45 jacks, which are terminated and then snapped into a protective faceplate or cover.

RJ-45 Modular Plug Termination.
RJ-45 modular plugs are commonly used to terminate four-pair twisted cable. These modular plugs are sometimes referred to as 8P8C (eight position eight contact). A VDV technician must select an RJ-45 modular plug that is compatible with the four-pair twisted cable involved. Compatibility factors include whether the cable is solid or stranded, the AWG gauge of the individual conductors, whether the cable is round or flat, and the Cat rating of the cable. Also, the RJ-45 modular plug must be compatible with the crimp tool. The following procedure can be used to terminate an RJ-45 modular plug:

1. Position the RJ-45 plug with the tab down, the gold contacts up and away, and the pin numbers showing 1 through 8, left to right.
2. Strip approximately 2″ of jacket from the cable.
3. Determine which termination pattern to use, either T568A or T568B.
4. Fan out, separate, and straighten the wires from the cable into the proper termination pattern.
5. Trim the wires at 90° and strip the insulation as specified by the RJ-45 plug OEM.
6. Fully insert the wires into the RJ-45 plug.
7. Insert the RJ-45 plug into the crimp tool, making sure that the wires stay in place, and crimp. **See Figure 9-24.**

RJ-45 Modular Plug Termination

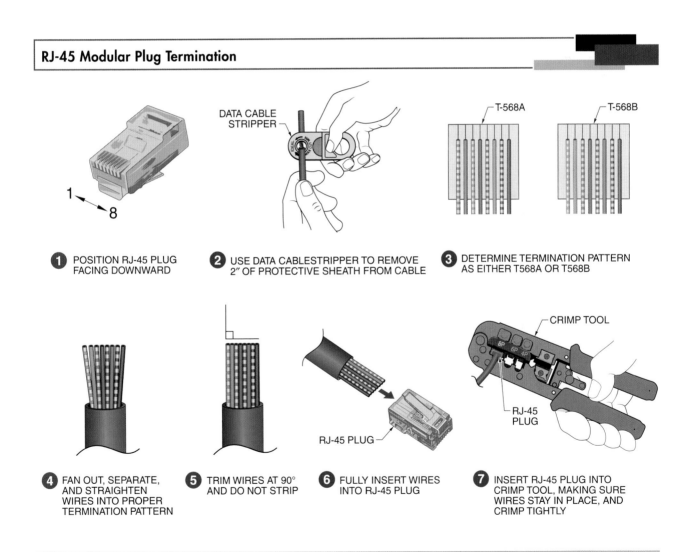

1 POSITION RJ-45 PLUG FACING DOWNWARD

2 USE DATA CABLESTRIPPER TO REMOVE 2″ OF PROTECTIVE SHEATH FROM CABLE

3 DETERMINE TERMINATION PATTERN AS EITHER T568A OR T568B

4 FAN OUT, SEPARATE, AND STRAIGHTEN WIRES INTO PROPER TERMINATION PATTERN

5 TRIM WIRES AT 90° AND DO NOT STRIP

6 FULLY INSERT WIRES INTO RJ-45 PLUG

7 INSERT RJ-45 PLUG INTO CRIMP TOOL, MAKING SURE WIRES STAY IN PLACE, AND CRIMP TIGHTLY

Figure 9-24. RJ-45 modular plugs are commonly used to terminate four-pair twisted cable.

Coaxial Cable Termination

Coaxial cable consists of an 18 AWG to 22 AWG center conductor, made of copper or copper-clad steel and surrounded by foam insulation. The foam insulation is commonly surrounded by a metallic foil, and the foil is surrounded by an aluminum or copper metallic braid. An outer jacket surrounds the metallic braid.

The most common types of coaxial cable are RG-6 and RG-59. RG-6 is used for present-day cable television (CATV) and satellite television installations. RG-59 is used for security equipment, surveillance equipment (CCTV), and older CATV installations.

The two most common types of coaxial connectors are 75 Ω (F-type) connectors and BNC-type connectors. F-type connectors are used for CATV and satellite television. BNC-type connectors are used for video applications, such as those for security and surveillance.

Splitters are used to distribute incoming CATV or satellite television signals. As a signal is "split," signal strength is lost. A *terminator* is a device installed on the end of a VDV line that absorbs a signal to prevent signal leakage through an unused port. Any unused ports on a splitter must have terminators installed for optimum performance. **See Figure 9-25.**

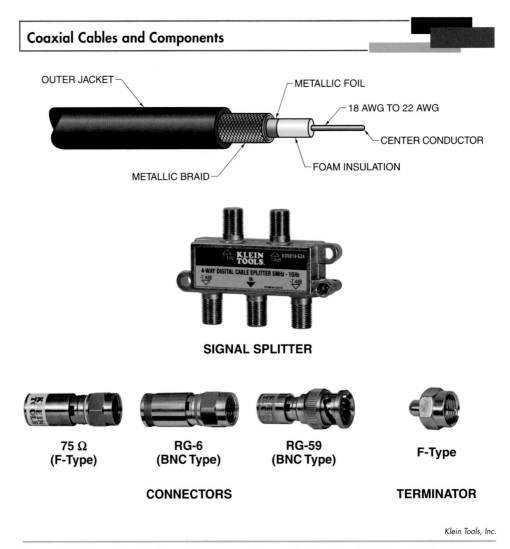

Coaxial Cables and Components

OUTER JACKET

METALLIC FOIL

18 AWG TO 22 AWG

CENTER CONDUCTOR

FOAM INSULATION

METALLIC BRAID

SIGNAL SPLITTER

75 Ω
(F-Type)

RG-6
(BNC Type)

RG-59
(BNC Type)

F-Type

CONNECTORS

TERMINATOR

Klein Tools, Inc.

Figure 9-25. Coaxial cable is used for present-day cable television (CATV), satellite television, security equipment, and surveillance equipment installations.

The tools required to terminate a crimp or compression-type coaxial connector are a cable stripper, a cable cutter, an electrician's scissors, and a crimper. Two types of strippers are multipurpose strippers and multistep strippers. Multipurpose strippers can be used to strip twisted-pair and coaxial cable. When used with coaxial cable, however, a multipurpose stripper must perform two or three separate stripping operations to prepare the cable for termination. A multistep stripper performs either two or three stripping operations simultaneously and thus saves time. Multistep strippers are only used with coaxial cables. The coaxial crimper must be compatible with the type of connector, F or BNC, and the specific OEM brand of connector.

The following procedure can be used to terminate an F-type connector:

1. Cut the end of the coaxial cable evenly.

2. Strip the jacket from the cable using a multipurpose stripper or multistep stripper per the instructions provided with the connector.

3. Fold the braid back over the jacket. Do not remove the metallic foil that surrounds the foam insulation.

4. Insert the center conductor into the inner barrel of the F-type connector by pushing the foam (surrounded by the foil) into the inner barrel until the foam is flush with the interior of the connector as viewed from the front of the connector. The center conductor should be approximately ⅛″ above the threaded collar of the connector. The outer barrel of the connector should cover the braid. Some connectors have a "push-out pin", which will fall out of the connector end when the proper depth is reached.

5. Insert the connector in the crimper and crimp per the OEM's instructions. **See Figure 9-26.**

F-Type Connector Termination Procedure

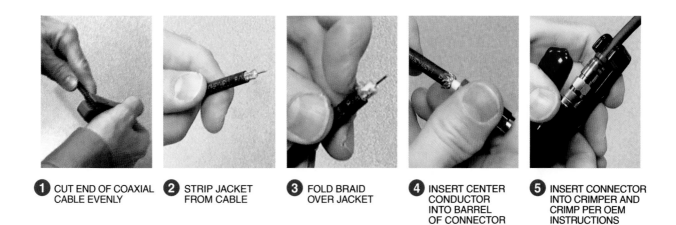

1 CUT END OF COAXIAL CABLE EVENLY   **2** STRIP JACKET FROM CABLE   **3** FOLD BRAID OVER JACKET   **4** INSERT CENTER CONDUCTOR INTO BARREL OF CONNECTOR   **5** INSERT CONNECTOR INTO CRIMPER AND CRIMP PER OEM INSTRUCTIONS

Figure 9-26. The tools required to terminate a crimp or compression-type coaxial connector to an F-type connector are a cable stripper, a cable cutter, an electrician's scissors, and a crimper.

Note: Always read and follow connector OEM instructions for exact dimensions, procedures, and tools for each connector type. The following procedure can be used to terminate a BNC connector:

1. Cut the end of the coaxial cable evenly.
2. Strip the cable using a multipurpose or multistep stripper per the OEM instructions. Do not remove the metallic foil that surrounds the foam insulation.
3. Slide the outer ferrule onto the cable and away from the cut cable end.
4. Place the center contact on the center conductor. Verify that it fits squarely against the foam insulation and that the metallic foil is not touching the center contact.
5. Crimp the center contact to the center conductor.
6. Insert the center contact into the connector body until it seats. The foam insulation should slide into the inner ferrule, and the braid should slide over the inner ferrule.
7. Slide the outer ferrule over the braid until it fits squarely against the connector body.
8. Insert the connector in the crimper and crimp per the OEM's instructions. **See Figure 9-27.**

Note: Always read and follow connector OEM instructions for exact dimensions, procedures, and tools for each connector type. The most current OEM instructions are typically found on the OEMs website rather than as hard copies.

BNC-Type Connector Termination Procedures

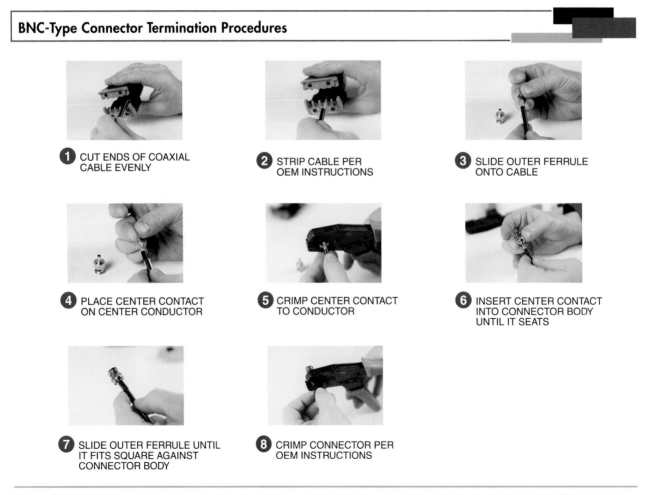

1. CUT ENDS OF COAXIAL CABLE EVENLY
2. STRIP CABLE PER OEM INSTRUCTIONS
3. SLIDE OUTER FERRULE ONTO CABLE
4. PLACE CENTER CONTACT ON CENTER CONDUCTOR
5. CRIMP CENTER CONTACT TO CONDUCTOR
6. INSERT CENTER CONTACT INTO CONNECTOR BODY UNTIL IT SEATS
7. SLIDE OUTER FERRULE UNTIL IT FITS SQUARE AGAINST CONNECTOR BODY
8. CRIMP CONNECTOR PER OEM INSTRUCTIONS

Figure 9-27. BNC-type connectors are terminated to coaxial cable in a procedure similar to that used for F-type connector terminations.

Summary

VDV technicians spend most of their time installing and terminating copper cabling. The two primary types of copper cable used in these systems are four-pair twisted cable and coaxial cable, with four-pair twisted cable being the most prevalent. The three steps involved with the installation and termination of copper cabling are rough-in, cable installation, and cable termination. The installation and termination of copper cabling is governed by codes, standards, specifications, and the requirements of cable and related equipment OEMs. VDV technicians must follow cable and termination device instructions provided by the OEMs to ensure VDV systems function properly.

Chapter Review

1. What are the three steps a VDV technician must follow to install and terminate a copper VDV system?

2. What material is used for backboards in TRs and ERs?

3. What devices are used to manage and route overhead cables entering TRs and ERs?

4. What are the three steps of cable installation?

5. As permitted by standards, what is the maximum length of a permanent link?

6. When pulling cable according to handling guidelines, what should be the maximum tension on the cable?

7. List nine different types of firestopping materials.

8. What is the most common termination pattern for four-pair twisted cable used in commercial installations?

9. What is the main function of a 110 connecting block?

10. What are the most common types of coaxial cables and their applications?

Chapter Review

11. When preparing a job site for rough-in procedures, what must be referenced to determine the proper placement of cables, receptacles, and related equipment?

12. What term does the International Building Code use when referring to fire-rated plywood?

13. What is a permanent link?

14. Why must a VDV technician track the starting length of the cable in each box or reel supplied by an OEM?

15. List three common sources of EMI.

16. List five factors that must be considered when pulling cable through a cable tray.

17. Explain the difference between a through penetration and a membrane penetration.

18. What are the two predominant conductor termination patterns for RJ-45 connectors?

19. What does twisted-pair cable management involve?

20. Explain how IDC termination works.

21. What is the most common tool used with IDC terminations?

22. How many wire positions are in a 110 block?

23. What are the two termination methods for 110-type RJ-45 jacks?

Chapter Review

24. What is a terminator?

25. Explain the differences between a telecommunications room and an equipment room.

Chapter Activity Cable Tray Print

Refer to the cable tray print section and answer the following questions.

CABLE TRAY LOAD AND FILL CHART

Flextray Series		Support Span/Loading Capacity*				Cable Fill (50% ‖‖)†		
Part Number	Part Number	5'-0"	6'-0"	7'-0"	8'-0"	Actual Area Inside Trays (in²)	Number of CAT 5e Cables‡	Number of CAT 6e Cables‡
FT1.5×12	1½" × 12"	29	17	14	11	12.2	176	124
FT 2×2	2" × 2"	34	28	24	20	4.3	61	43
FT 2×4	2" × 4"	52	43	35	27	8.2	118	83
FT 2×5	2" × 6"	66	47	35	27	12.1	175	123
FT 2×8	2" × 8"	66	47	35	27	16.1	231	163
FT 2×12	2" × 12"	68	47	35	27	23.9	345	243
FT 2×16	2" × 16"	68	47	35	27	31.8	459	324
FT 2×18	2" ×1 8"	68	47	35	27	35.8	516	364
FT 2×20	2" × 20"	68	47	35	27	39.7	573	404
FT 2×24	2" × 24"	68	47	35	27	47.5	686	484
FT 2×30	2" × 30"	68	47	35	27	59.8	862	608
FT 2×32	2" × 32"	77	53	39	30	63.3	914	645
FT 4×4	4" × 4"	58	49	42	36	15.8	227	160
FT 4×6	4" × 6"	93	77	60	46	23.6	341	240
FT 4×8	4" × 8"	94	78	61	47	31.5	454	321
FT 4×12	4" × 12"	119	83	61	47	47.5	686	484
FT 4×16	4" × 16"	119	83	61	47	63.5	917	647
FT 4×18	4" × 18"	119	83	61	47	71.5	103.2	728
FT 4×20	4" × 20"	119	83	61	47	79.5	1148	810
FT 4×24	4" × 24"	128	89	65	50	95.5	1379	973
FT 4×30	4" × 30"	128	89	65	50	119.5	1725	1217
FT 6×6	6" × 8"	111	77	57	43	47.3	682	481
FT 6×12	6" × 12"	124	86	63	48	71.6	1034	729
FT 6×16	6" × 16"	128	89	65	50	95.3	1375	970
FT 6×18	6" × 18"	128	89	65	50	107.3	1549	1092
FT 6×20	6" × 20"	141	98	72	55	118.9	1716	1211
FT 6×24	6" × 24"	154	107	78	60	143.3	2068	1459

* Published load chart has not been tested with Flexmate splice. Please consult the factory for load information when using the flexmate option.

† Flextray fill load is based on NEC allowable fill of 50%. The NEC rule requires that the cable cross-sectional areas together may not exceed 50% of the tray area (width x depth = fill). Cables will nearly completely fill the cable tray when reaching the 50% cable fill, due to empty space between the surface of the cables. TIA recommends 40% fill ratio. Flextray loads shown in the loading chart will not be exceeded at 50% fill.

‡ CAT 5e 4-pr non-plenum approximated at 0.21 in. diameter, CAT 6 4-pr non-plenum approximated at 0.25 in. diameter. Actual diameters vary by cable manufacturer.

SHEET NOTES

1. PROVIDE WIRE MESH CABLE SUPPORT SYSTEM FOR DISTRIBUTION OF AUDIOVISUAL AND TELECOMMUNICATIONS CABLE.

2. CABLE TRAY SHALL B-LINE FLEXTRAY OR EQUAL. SIZE OF THE CABLE TRAY AS NOTED IN CABLE TRAY LEGEND.

3. INSTALLATION CONTRACTOR SHALL COORDINATE WITH ARCHITECTURAL, PLUMBING AND HVAC PLANS EXACT LOCATION OF CEILING MOUNTED DEVICES PRIOR OF PLACEMENT CABLE TRAY.

4. COORDINATE ARCHITECT COLOR OF THE CABLE TRAY AND CABLE TRAY ACCESSORIES. STANDARD FINISH IS GALVANIZED ZINC.

5. FOR MAXIMUM CABLE TRAY LOAD CAPACITY INSTALL CENTER HUNG SUPPORT KIT OR TRAPEZE SUPPORT KIT EVERY 5-FEET.

6. INSTALL CABLE DROP-OUT MAINTAIN CABLE BEND RADIUS AT THE TRANSITION OR EXIT POINTS.

7. PROVIDE BARRIER KIT FOR CABLE SEPARATION WITHIN CABLE TRAY.

8. PROVIDE TO A/V INFRASTRUCTURE PLAN, SHEET AV-101A FOR CONDUIT STUBS SIZE REQUIREMENTS AND MOUNTING HEIGHT OF A/V DEVICES.

9. PROVIDE CONDUIT CONNECTOR AND CONDUIT STUBS TO THE CABLE TRAY FROM A/V AND TELECOM OUTLETS.

10. REFER TO A/V INFRASTRUCTURE PLAN, SHEET AV-101A FOR CONDUIT STUBS SIZE REQUIREMENTS AND MOUNTING HEIGHT OF A/V DEVICES.

11. REFER TO CABLE LOCATION PLAN, SHEET T-1.0.0 FOR CONDUIT STUBS SIZE REQUIREMENTS AND MOUNTING HEIGHT OF TELECOM OUTLETS.

KEY NOTES

1. WAP MOUNTING HEIGHT AT 10'- 00" A.F.F

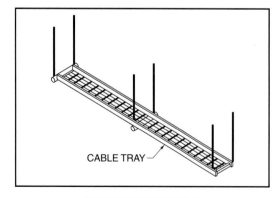

CABLE TRAY

CABLE TRAY DETAIL

Conduit Legend	
	½" Chase
	2" Chase
	½" EMT
	¾" EMT
	1" Chase
	1½" EMT

Chapter Activity Cable Tray Print

1. What is the maximum allowable cable fill percentage per the NEC®?

2. What is the maximum allowable cable fill percentage per the TIA?

3. What is the approximate diameter of a Cat 5e four-pair nonplenum cable?

4. What is the approximate diameter of a Cat 6 four-pair nonplenum cable?

5. What is the maximum number of Cat 5e four-pair nonplenum cables allowed in a 4″ × 18″ cable tray?

6. What is the maximum number of Cat 6 four-pair nonplenum cables allowed in a 4″ × 18″ cable tray?

7. Provide the part number and size of cable tray required for an installation of 500 Cat 5e four-pair nonplenum cables.

8. What is the maximum loading capacity (lb/ft) of the cable tray selected in question 7 with a support span of 7′-0″?

9. What support span is required per the prints?

Chapter 10

Testing Copper VDV Systems

Testing a copper system is an important part of a VDV technician's job. It is used to troubleshoot VDV systems and is the final step in installing a new VDV system. Testing is typically divided into three categories. The categories in the order of least to most complex are verification testing, qualification testing, and certification testing. Verification testing identifies cables and is used to confirm that the physical infrastructure, cables, and termination devices are wired correctly. Qualification testing is used to confirm that the physical infrastructure can support a specified network technology. Certification testing is used to confirm that the physical infrastructure complies with Telecommunications Industry Association (TIA) or International Organization for Standardization (ISO) standards. Structured cabling system OEMs require certification in order to provide a warranty for an installation.

OBJECTIVES

- Explain the differences between verification, qualification, and certification testers.
- Describe how toners and wiremap testers are used to test copper VDV systems.
- Explain the differences between qualification tests and diagnostic tests.
- Describe certification tests and certification test parameters.

Digital Resources

ATPeResources.com/QuickLinks
Access Code 838502

VERIFICATION TESTERS

A *verification tester* is a test instrument designed to trace the path of conductors, and confirm that conductors and data outlets are wired correctly. When troubleshooting a cable to determine the cause of a problem, the first test instrument a VDV technician typically uses after a visual inspection is a verification tester. These testers are the most common test instruments used by VDV technicians. They are also the least complex and least expensive. The two most common types of verification testers are toners and wiremap testers. **See Figure 10-1.**

Verification Testers

TONER

WIREMAP TESTER

Fluke Networks

Figure 10-1. The two most common types of verification testers are toners and wiremap testers.

Toners

A *toner* is a test instrument designed to identify and locate a copper conductor or cable. It can also be used to test for breaks and other problems, by sending a signal through the conductor or cable. Toners can be used with individual conductors, twisted-pair cables, and coaxial cables.

A toner consists of two separate parts, a tone generator and a toner probe. The tone generator is the part of the toner connected to the conductors using integral test leads with alligator clips or an integral test cable with an RJ-11 plug. The tone generator produces a signal, either continuous or alternating, in conductors. The toner probe has an inductive probe, an amplifier, and a speaker with volume control. The toner probe is also known as an inductive amplifier. It produces an audible sound when it detects a signal from a tone generator.

Toners are intended for use on conductors and cables that do not have any AC voltage, DC voltage, or active VDV signals. Before using a specific toner, VDV technicians must read and follow the OEMs instructions for the toner. VDV technicians must also follow all safety precautions listed by the OEM, including using the required personal protective equipment (PPE).

Identifying Cables with Toners. Toners are frequently used to identify a specific coaxial or twisted-pair cable, which is unmarked, behind a wall, under a floor, in a ceiling, or part of a large cable bundle. The tone generator is connected to one end of the cable. The toner probe is then used to identify the opposite end. **See Figure 10-2.** To identify a cable using a toner, apply the following procedure:

1. Connect the red lead of the tone generator to a wire of a twisted-pair cable or to the metallic shield of a coaxial cable.

2. Connect the black lead of the tone generator to another wire (preferably not of the same pair) of the twisted-pair cable or to a ground if available. For coaxial cable, connect the black lead to the center conductor of the cable or to a ground if available.

3. Turn the tone generator on.

4. Turn the toner probe on.

5. Move the toner probe tip along the surface of each cable, along the surface of the wall adjacent to the cable installation, or along and around the circumference of the cable bundle. When testing cables or bundles, the cable that produces the loudest tone is the cable being tested. When testing along a wall, the cable is behind the wall in the area where the loudest tone is heard. *Note:* The signal from the tone generator may spill over to adjacent cables if they are tightly bundled. Cables in a bundle should be separated at least 1″ apart in order to identify the cable being tested.

6. Remove the tone generator leads from the cable wires.

7. Turn the tone generator and toner probe off.

Identifying Wire Pairs with Toners. Toners are also used to identify a specific pair of wires in a multiconductor cable. The tone generator is connected to a pair of wires at one end of the cable. The toner probe is then used to identify the pair at the opposite end. **See Figure 10-3.** To identify a wire pair using a toner, apply the following procedure:

1. Connect the red lead of the tone generator to one wire of the twisted pair.

2. Connect the black lead of the tone generator to the second wire of the twisted pair.

3. Turn the tone generator on.

4. Turn the toner probe on.

5. Move the toner probe tip along and around the circumference of the cable pairs at the opposite end of the cable. The pair that produces the loudest tone is the pair being tested.

6. Separate the pair being tested from the other pairs and untwist the wires. Place the toner probe tip near one wire of the pair. *Note:* The tone level should be loud.

7. Move the toner probe tip toward the second wire of the pair. The tone level should decrease as the toner probe tip moves away from the first conductor and increase as it nears the second conductor.

8. Remove the tone generator leads from the wire.

9. Turn the tone generator and toner probe off.

Identifying Cables with Toners

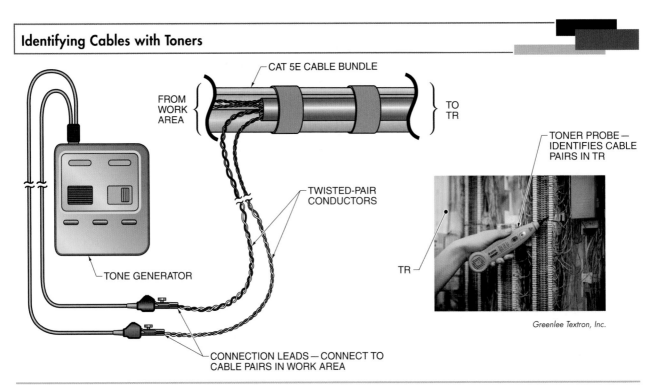

Figure 10-2. Toners are frequently used to identify a specific coaxial or twisted-pair cable, which is unmarked, behind a wall, under a floor, in a ceiling, or part of a large cable bundle.

Identifying Wire Pairs with Toners

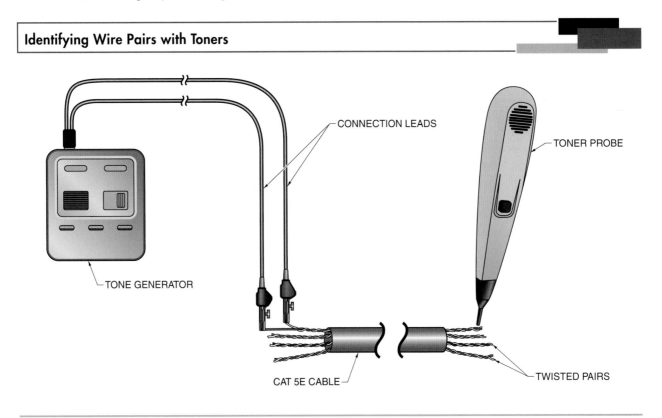

Figure 10-3. Toners are also used to identify a specific pair of wires in a multiconductor cable. The tone generator is connected to a pair of wires at one end of the cable, and the toner probe is used to identify the pair at the opposite end.

Identifying Work Area Outlets with Toners. In addition to identifying individual cables and wire pairs, toners are used to identify work area outlets such as RJ-45 jacks. The tone generator is connected to a work area outlet via a test cable terminated with an RJ-45 plug. The toner probe is then used to find the corresponding outlet at the patch panel. **See Figure 10-4.** To identify work area outlets using a toner, apply the following procedure:

1. Insert the RJ-45 plug of the tone generator into the work area outlet.

2. Turn the tone generator on.

3. Go to the telecommunications room (TR) or equipment room (ER) where the work area outlet terminates and turn the toner probe on.

4. Move the toner probe tip along the RJ-45 jacks on the patch panel where the outlets terminate. The jack that produces the loudest tone is the corresponding end of the work area outlet being tested. *Note:* The signal from the tone generator may leak into adjacent RJ-45 jacks, making it difficult to find the correct jack. When this occurs, connect a short patch cord into the affected jack and move the toner probe along the patch cord. The short patch cord allows the VDV technician to determine which jack has the loudest tone.

5. Turn the toner probe off.

6. Remove the RJ-45 plug of the tone generator from the work area outlet.

7. Turn the tone generator off.

Identifying Work Area Outlets with Toners

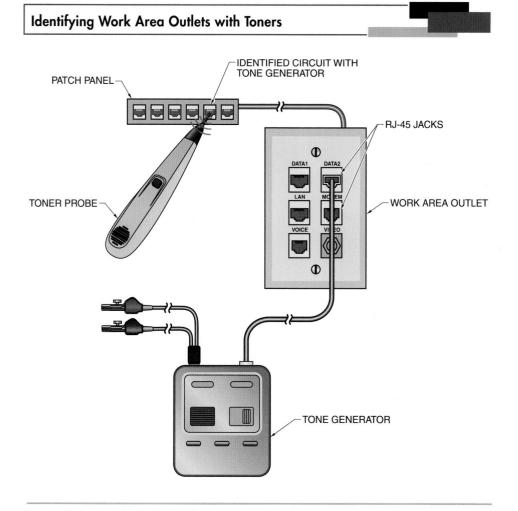

Figure 10-4. In addition to identifying individual cables and cable pairs, toners are used to identify work area outlets such as RJ-45 jacks.

Wiremap Testers

Wiremap testers are used to verify that cables and VDV receptacles are wired correctly. They can be used with twisted-pair cables (unshielded or shielded) and coaxial cables. A wiremap tester consists of two separate elements, a wire mapper and a remote unit. The wire mapper consists of navigation keys and buttons. The remote unit is required to complete the wiremap test. A remote unit usually has a set of lights that blink or a speaker that emits a sound to indicate if a system is wired correctly and to help a VDV technician identify twisted pairs or cables. The wire mapper connects to a cable or a VDV receptacle with a patch cord. The remote unit connects to the corresponding end of the cable or receptacle (RJ or coaxial) with a patch cord. More than one remote unit can be used when testing multiple outlets. When more than one remote unit is used, they are called numbered remote identifier units. Wiremap testers are also referred to as cable mappers or cable testers. **See Figure 10-5.**

Wiremap Testers

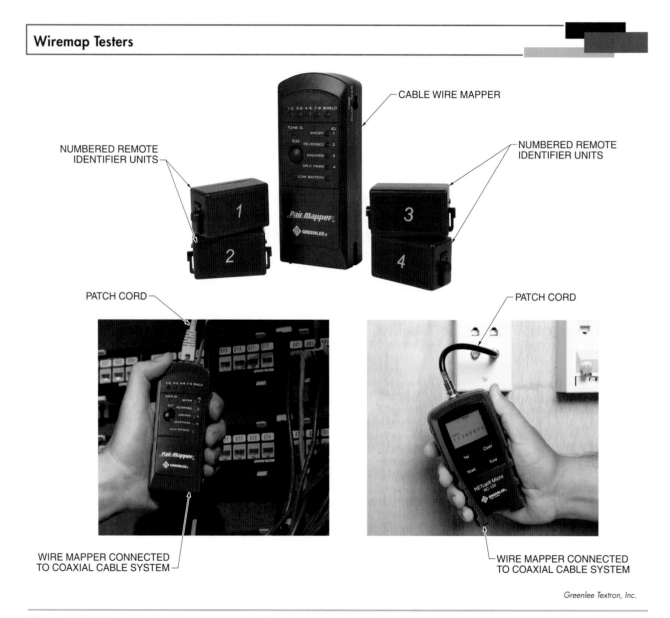

Greenlee Textron, Inc.

Figure 10-5. Wiremap testers are used to verify that cables and VDV receptacles are wired correctly and can be used on twisted pair cable or coaxial cable.

Wiremap testers are used to determine correct wiring, open pairs, shorted pairs (short circuits), crossed pairs, crossed wires, split pairs, the distance to faults, and cable lengths for twisted-pair cables. They can also identify data outlets, identify active Ethernet ports, identify PoE (power over Ethernet) ports, and serve as tone generators for twisted-pair cables. These testers are also used to determine correct wiring, opens, short circuits, the distance to faults, and cable lengths for coaxial cables as well as identify coaxial outlets and serve as tone generators.

Wiremap testers are not intended for use on active telephone systems, active telephone equipment, or AC circuits. Before using a wiremap tester, VDV technicians must read and follow the wiremap tester OEM's instructions. VDV technicians must also follow all safety rules, including using the required PPE.

Tech Tip

Power over Ethernet (PoE) is a method of supplying power to remotely located electronic devices. Video cameras, wireless access points (WAPs), and voice over Internet protocol (VoIP) telephones commonly use PoE. These devices require power to operate but are often placed in locations where power is not readily available. Without PoE, a separate cable run would be required to supply power. Because standard Ethernet transmission only requires the use of two of the pairs in a Cat cable, the other two pairs can be used for power. In addition, as most wiring closets have backup power supplies, the devices also continue to operate when power failures occur. The original standard was for up to 15.4 W of power (IEEE 802.3af in 2003). This was later updated to 25.5 W (IEEE 802.3at in 2009). OEMs are researching PoE for devices that would require a greater amount of power. Examples include LED lighting systems and outdoor (heated) PTZ cameras.

Verifying Twisted-Pair Cables and Data Outlets with Wiremap Testers. One of the most common uses for a wiremap tester is to verify that twisted-pair cables and data outlets are wired correctly. The indicators of the wire mapper provide a visual indication of the test results. If there is a problem with the wiring, the wire mapper displays the nature of the problem and if the tester model has this feature, the distance to it. This information allows the VDV technician to quickly locate the problem. If the wiring is correct, the wire mapper displays the correct wiring and the total distance from the mapper to the remote unit, including any patch cords. This information allows the VDV technician to verify that the cable length does not exceed the maximum allowed by standards or specifications. **See Figure 10-6.** To verify that a twisted-pair cable and its data outlets are wired correctly using a wiremap tester, apply the following procedure:

1. Connect the remote unit to a cable or data outlet.
2. Connect the wire mapper to the other end of the cable or data outlet.
3. Turn the wire mapper on and view the results.
4. If the wire mapper indicates a problem, perform the required procedures to correct the problem and retest. Repeat this step until the test results are satisfactory.
5. Turn the wire mapper off.
6. Disconnect the wire mapper from the cable or data outlet.
7. Disconnect the remote unit from the other end of the cable or data outlet.

Twisted-Pair Cable Faults. There are many types of twisted-pair cable faults. However, VDV technicians commonly use wire mappers to test for open pairs, shorted pairs, crossed wires, crossed pairs, and split pairs.

An *open pair* is a fault that occurs in VDV cabling when one or both conductors of a pair are not connected (open). The pair does not have electrical continuity from end to end because of the fault. A *shorted pair* is a fault that occurs when the two conductors of a VDV cable pair are short-circuited and continue to have electrical continuity. This condition typically occurs when a conductor from one pair is short-circuited to a conductor of another pair.

Verifying Twisted-Pair Cables and Data Outlets with Wiremap Testers

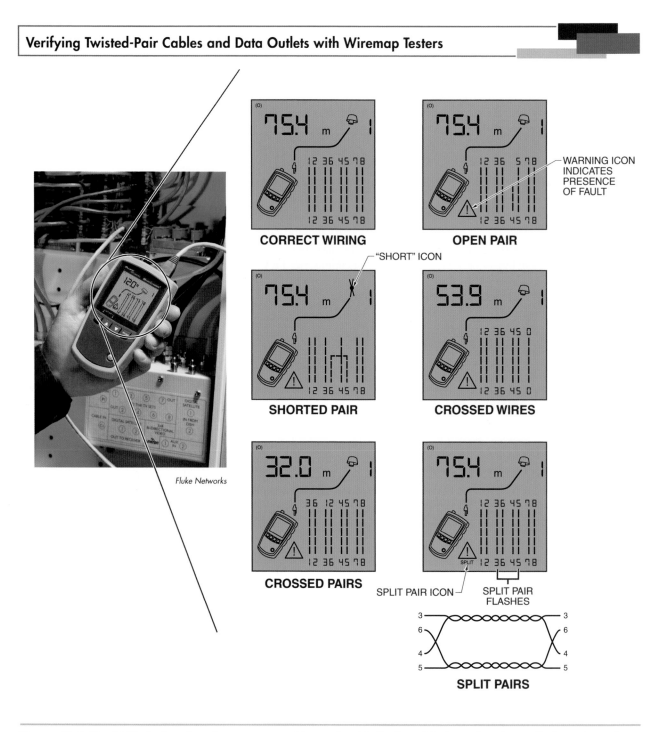

Fluke Networks

Figure 10-6. The LCD display of a wire mapper provides a visual indication of test results, along with the nature of any problems and the distance to them.

A *crossed wire* is a fault that occurs when the termination positions of two VDV conductors are transposed, or connected to the wrong conductor, at one end of a cable. A *crossed pair* is a fault that occurs when the termination positions of two pairs are transposed, or connected to the wrong pair, at one end of a VDV cable. A *split pair* is a fault that occurs in a VDV cable when the wire in one pair is interchanged with a wire in another pair.

Some faults are caused by damage to the cable, but most are caused by improper terminations. Upon finding a fault, VDV technicians typically reterminate the cable and retest it until a satisfactory result is achieved.

Identifying Data Outlets with Wiremap Testers. In addition to verifying that twisted-pair cables and data outlets are wired correctly, a wiremap tester can also identify data outlets. **See Figure 10-7.** When performing this procedure, numbered remote identifier units are often used. Numbered remote identifier units each have a unique number that allows a VDV technician to easily distinguish them while testing multiple outlets at once. To identify data outlets using a wire mapper with numbered remote identifier units, apply the following procedure:

1. Connect each numbered remote identifier unit to a work area data outlet.

2. Connect the wire mapper to the patch panel for the outlets.

3. Turn the wire mapper on and view the test results on the display screen.

4. If there is a problem, perform the required procedures to correct it and retest. Repeat this step until the test results are satisfactory.

5. Identify each work area outlet and patch panel outlet and record the information.

6. Label the work area outlets and patch panel outlets per the standards or job specifications.

7. Turn the wire mapper off.

8. Disconnect the wire mapper from the patch panel.

9. Disconnect the numbered remote identifier units from the work area data outlets.

Identifying Data Outlets with Wiremap Testers

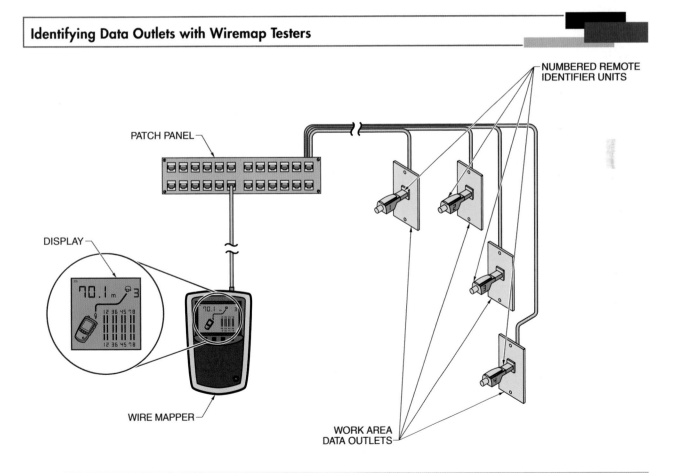

Figure 10-7. In addition to verifying that twisted-pair cables and data outlets are wired correctly, a wiremap tester can also identify data outlets.

Verifying Coaxial Cables and Outlets with Wiremap Testers. Sometimes, VDV technicians use wiremap testers to verify that coaxial cables and coaxial outlets are wired correctly. The LCD display of the wire mapper provides a visual indication of the test results. If there is a problem, the wire mapper displays the problem and the distance to it. This information allows the VDV technician to quickly locate the problem. If no problems are detected, the wire mapper displays the correct wiring and the distance from the wire mapper to the remote unit, including any patch cords. When testing a cable, this information allows the VDV technician to verify that the cable length does not exceed the maximum allowed by the standards or specifications. **See Figure 10-8.** To verify that a coaxial cable or coaxial outlet is wired correctly using a wiremap tester, apply the following procedure:

Verifying Coaxial Cables and Outlets with Wiremap Testers

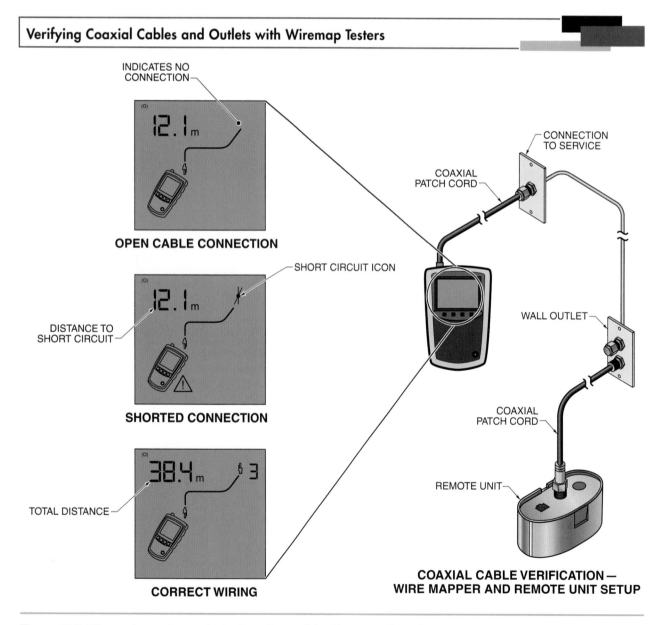

INDICATES NO CONNECTION

OPEN CABLE CONNECTION

SHORT CIRCUIT ICON

DISTANCE TO SHORT CIRCUIT

SHORTED CONNECTION

TOTAL DISTANCE

CORRECT WIRING

CONNECTION TO SERVICE

COAXIAL PATCH CORD

WALL OUTLET

COAXIAL PATCH CORD

REMOTE UNIT

COAXIAL CABLE VERIFICATION — WIRE MAPPER AND REMOTE UNIT SETUP

Figure 10-8. When using a wiremap tester to verify coaxial cable connections, the wire mapper displays the problem, if one exists, and the distance to it.

1. Connect the remote unit to the coaxial cable or outlet.

2. Connect the wire mapper to the corresponding end of the cable or outlet.

3. Turn the wire mapper on and view the test results on the wire mapper display screen.

4. If there is a problem, perform the required procedures to correct it and retest. Repeat this step until the test results are satisfactory.

5. Turn the wire mapper off.

6. Disconnect the wire mapper from the end of the cable or outlet.

7. Disconnect the remote unit from the corresponding end of the cable or outlet.

Coaxial Cable Faults. The two most common types of coaxial cable faults are open coaxial cables and shorted coaxial cables. These faults are similar to the faults found with twisted-pair cables.

With an open coaxial cable, one or both conductors of the cable are open. The shield, center conductor, or both do not have electrical continuity from end to end because of the open connection. With a shorted coaxial cable, the two conductors of the cable are short-circuited. The shield and center conductor are unintentionally connected and have electrical continuity.

Some faults to coaxial cable are caused by damage to the cable, but most are caused by improper terminations. Upon finding a fault, VDV technicians typically reterminate the cable and retest it until a satisfactory result is achieved.

Identifying Coaxial Data Outlets with Wiremap Testers. In addition to verifying that coaxial cables and outlets are wired correctly, a wiremap tester can also identify coaxial data outlets. **See Figure 10-9.** When performing this procedure, numbered remote identifier units are often used. To identify coaxial data outlets using a wiremap tester, apply the following procedure:

1. Connect the numbered remote identifier units to the work area coaxial data outlets.

2. Connect the wire mapper to the coaxial patch panel.

3. Turn the wire mapper on and view the test results on the display screen.

4. If there is a problem, perform the required procedures to correct it and retest. Repeat this step until the test results are satisfactory.

5. Identify each work area outlet and patch panel outlet and record the information.

6. Label the work area outlets and patch panel outlets per the standards or job specifications.

7. Turn the wire mapper off.

8. Disconnect the wire mapper from the patch panel.

9. Disconnect the numbered remote identifier units from the work area coaxial data outlets.

Identifying Coaxial Data Outlets with Wiremap Testers

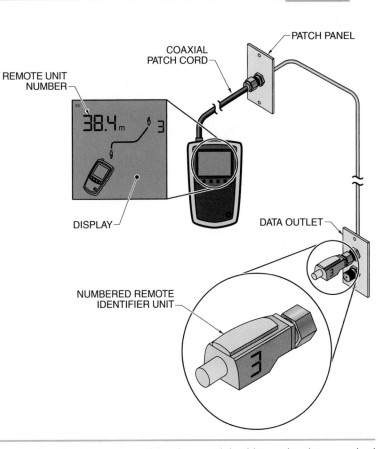

Figure 10-9. In addition to verifying that coaxial cables and outlets are wired correctly, a wiremap tester can also identify coaxial data outlets.

QUALIFICATION TESTERS

A *qualification tester* is a test instrument designed to determine if a VDV cable has a connectivity problem and if the cable can support the bandwidth requirements of a specific network. A qualification test is comprised of several separate tests, conducted simultaneously by a VDV technician using a qualification tester. A "pass" or "fail" result is based on cable length, the wiremap, and signal performance. Test results can be electronically recorded and transferred to a computer. The tests performed with a qualification tester are more complex than those performed with verification testers.

Qualification testers can be used on twisted-pair cable (unshielded or shielded) or coaxial cable. A qualification tester consists of two separate elements–a qualification unit and a remote unit. The qualification unit has an LCD display, navigation keys, and buttons. **See Figure 10-10.** The qualification unit connects to a cable or to a data outlet with a patch cable. The remote unit connects to the corresponding end of the cable or outlet with a patch cord. The remote unit is required for a complete series of tests. As with wiremap testers, numbered remote identifier units can be used when testing multiple outlets. Qualification testers are sometimes referred to as network testers.

Qualification Testers

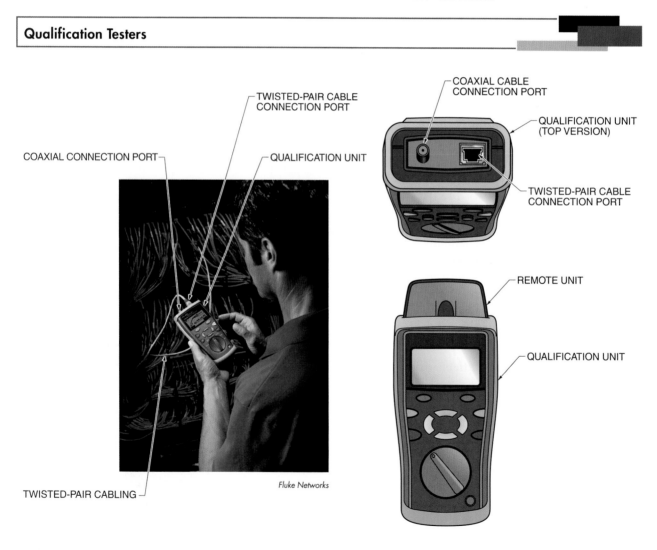

Fluke Networks

Figure 10-10. A qualification tester consists of two separate elements—a qualification unit and a remote unit.

In addition, qualification testers can identify data outlets and coaxial outlets using numbered remote identifier units, serve as tone generators, verify continuity, and identify ports, PC settings, and coaxial signal strengths. These tests are not performed on active telephone systems or active telephone equipment. A network must be inactive when undergoing a qualification test. Before using a specific qualification tester, VDV technicians must read and follow all instructions. VDV technicians must also follow all safety rules, including using the required PPE.

Qualification Tests

Qualification tests are performed to determine if a cable has sufficient bandwidth to support a particular network. Qualification testers provide a "pass" or "fail" result for the cable being tested. The results involving cable length, the wiremap, and signal performance are displayed on the LCD screen. If the cable fails, the tester provides data on the failure. Cabling for the following networks can be qualified: coaxial, 10Base-T, 100Base-TX, 100Base-T, 10GBase-T, VoIP, and analog telephone services. **See Figure 10-11.** To perform a test with a qualification tester, apply the following procedure:

1. Turn the qualification tester on.
2. Select the "qualification" function.
3. Connect the remote unit to the work area cable or data outlet to be tested.
4. Connect the qualification tester to the corresponding end of the cable or data outlet on the patch panel in the TR or ER.
5. Perform the qualification test. Save and view the results.
6. If the cable does not pass for the required bandwidth, view the detailed results.
7. If possible, correct the problem and retest. Repeat this step until the test results are satisfactory.
8. Disconnect the tester from the end of the cable or data outlet.
9. Disconnect the remote unit from the other end of the cable or data outlet.
10. Turn the qualification tester off.

Qualification Tests

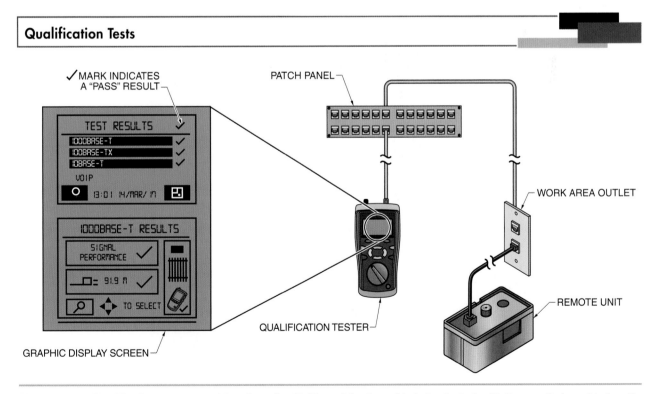

Figure 10-11. Qualification testers provide a "pass" or "fail" result for the cable being tested, with the results for cable length, the wiremap, and signal performance displayed on the LCD screen.

Qualification Tester Diagnostic Tests

Most qualification testers can also be used to perform diagnostic tests. The number and types of diagnostic tests available vary between models of testers. Many diagnostic tests are performed on active networks. **See Figure 10-12.** Common diagnostic tests include "blinking" a port, measuring signal strength, and performing a wiremap test.

"Blinking" the port of a hub or switch can help a VDV technician determine which port a cable is connected to. The qualification tester is connected to the work area data outlet, and a signal is sent to the network device's activity LED. This makes the LED flash, or blink. The test is useful when cable labeling is in question. Unlike a toner, removing the cables from their ports is not required to complete the test.

Qualification Tester Diagnostic Test

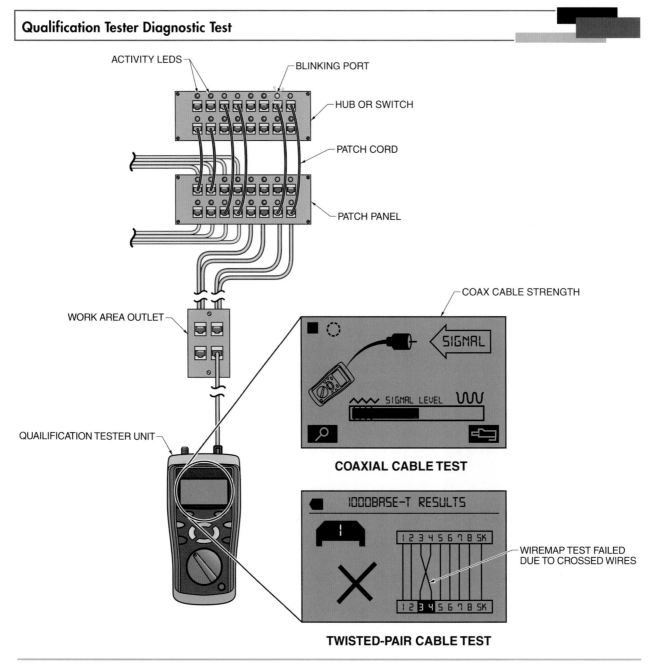

Figure 10-12. Qualification tester diagnostic tests include "blinking" a port, measuring signal strength, and wiremap tests.

Qualification testers can also measure the signal strength of devices connected to a coaxial cable. These types of devices may include CatVs, antennas, satellite dishes, and DVD players. It is also possible to obtain the speed of a hub or network interface card in a computer. To do this, the qualification tester is connected to one end of a twisted-pair cable to determine the speed of the attached port.

CERTIFICATION TESTERS

A *certification tester* is a test instrument used to test and certify cables in a system in order to obtain a warranty from the structured cabling system OEM. These testers are more complex than verification testers or qualification testers. *Note:* Only certification testers can be used to certify network cabling. Qualification and verification testers cannot be used for this purpose.

Certification testers are designed to measure the performance of a cable and associated components, such as patch panel jacks and work area jacks, to determine if their performance meets a standard. Most certification testers can test for either TIA standards or ISO standards. Typically, TIA standards are used in North America, and ISO standards are used everywhere else. In addition to determining the test standard, a VDV technician must determine the proper cable category (Cat 5, 5e, 6, or 6A).

The certification test itself is comprised of several separate tests conducted simultaneously. The tests performed vary depending on the standard and cable category selected. The certification tester provides a "pass" or "fail" result, and recording the result is required by the standards. Results can be saved on a computer or on a portable memory device, such as a USB flash drive. Downloading the test results to a computer or flash drive on a daily basis minimizes the risk of lost data.

A certification tester can be used with twisted-pair cables (unshielded or shielded), coaxial, or fiber-optic cables. A certification tester consists of two separate elements, a main tester and a remote tester. The main tester has an LCD display, navigation keys, and buttons. The remote tester has a limited number of buttons and LED indicators. The main tester connects to a data outlet through a patch cable, and the remote tester connects to the corresponding end of the outlet through a patch cord. Certification testers are also referred to as cable certifiers and cable analyzers.

Certification testers must be factory calibrated periodically in order to maintain the level of accuracy required for certification. Test results from a tester that does not have a current valid calibration are likely to be rejected by the end user. In addition, a certification tester can perform many of the functions of a qualification tester or a verification tester. Before using a specific certification tester, VDV technicians must read and follow all instructions. VDV technicians must also follow all safety rules, including using the required PPE.

Certification Tests

Certification tests are most commonly performed on a permanent link, since optimum performance on the permanent link is what a structured cabling OEM guarantees. A *link* is an end-to-end transmission path provided by a VDV cabling system. A permanent link involves the installed cable, the connectors, the cross-connects, and the outlets. A *cross-connect* is a facility enabling the termination of network cables as well as their interconnection or cross-connection with other cables or devices. (Cross-connects are also known as punchdowns.) In copper-based systems, patch cords are used to connect equipment to cross-connects. Permanent link testing includes the patch panel jack, the horizontal cabling, and the wall outlet jack. It covers the items installed by the VDV contractor and is required to obtain a manufacturer's warranty.

Certification tests can also be performed on a channel. A channel test involves the end-to-end transmission or communications path over which application-specific equipment is connected. Channel testing includes the patch cord that connects a device to the patch panel, the patch panel jack, the horizontal cabling, consolidation points, the wall outlet jack, and the patch cord that connects a device to the wall outlet jack. **See Figure 10-13.**

Certification testers provide "pass" or "fail" results. The results are displayed on the tester's display screen. If a tested link fails, the certification tester provides detailed information on the fault. The information provided may include the type of fault, the distance to the fault, and the possible corrective action. Certification testers should not be connected to active data or telephone networks. A network must be inactive when performing a certification test. To perform a certification test using a certification tester, apply the following procedure:

1. Turn the main tester and remote tester on.
2. Select the standard (TIA or ISO) and cable category.
3. Connect the remote tester to the work area outlet.
4. Connect the main tester to the corresponding end of the outlet on the patch panel in the TR or ER.
5. Perform the certification test. Save and view the results.
6. If the link does not pass, view the detailed fault information.
7. If possible, correct the problem and retest. Repeat this step until the test results are satisfactory.
8. Disconnect the main tester from the patch panel.
9. Disconnect the remote tester from the work area outlet.
10. Turn the main tester and remote tester off.

Certification Test Parameters

Certification test parameters are automatically selected depending on the standard and cable category involved. Typically, the higher the cable category number, the higher the required cable performance level and the greater the number of certification test parameters. Each individual test parameter has an acceptable range or level that must be met in order to pass. The tested link must pass all of the test parameters in order to receive a "pass" result from the certification tester.

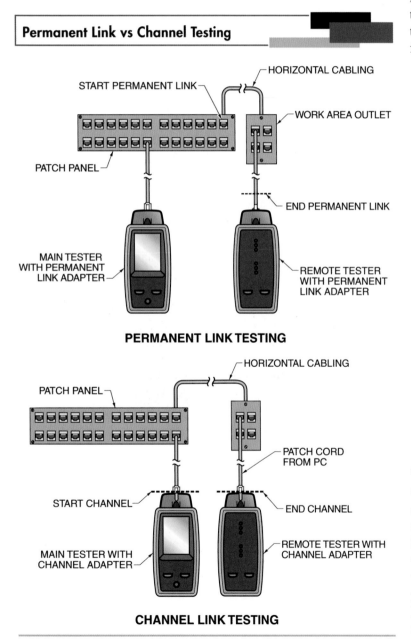

Permanent Link vs Channel Testing

HORIZONTAL CABLING
START PERMANENT LINK
WORK AREA OUTLET
PATCH PANEL
END PERMANENT LINK
MAIN TESTER WITH PERMANENT LINK ADAPTER
REMOTE TESTER WITH PERMANENT LINK ADAPTER

PERMANENT LINK TESTING

HORIZONTAL CABLING
PATCH PANEL
PATCH CORD FROM PC
START CHANNEL
END CHANNEL
MAIN TESTER WITH CHANNEL ADAPTER
REMOTE TESTER WITH CHANNEL ADAPTER

CHANNEL LINK TESTING

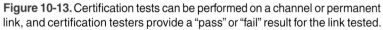

Figure 10-13. Certification tests can be performed on a channel or permanent link, and certification testers provide a "pass" or "fail" result for the link tested.

Typical certification test parameters include cable length, a wiremap, delay skew, insertion loss, attenuation, near-end crosstalk (NEXT), far-end crosstalk (FEXT), power sum near-end crosstalk (PS NEXT), attenuation to crosstalk ratio far-end (ACR-F), power sum attenuation to crosstalk ratio far-end (PS ACR-F), and return loss.

A certification tester measures the length of a cable to determine if it complies with the associated standard. Due to the lay of the twists in twisted-pair cables, the actual lengths of the twisted pairs can vary. When providing a wiremap test, a certification tester tests for open pairs, shorted pairs, crossed wires, crossed pairs, and split pairs.

Delay skew is the signal delay difference between the fastest and slowest pair in a twisted-pair cable. Delay skew is also known as propagation delay. *Insertion loss* is the loss or degradation of a signal in a VDV cable when a connector is inserted into another connector or associated hardware. *Attenuation* is the reduction in power of any signal, light beam, or light wave, either completely or as a percentage of a reference value. Attenuation is also known as the measure of signal lost in the cable length.

Near-end crosstalk (NEXT) is the measure of the amount of signal interference from one pair into the pair next to it in the same cable. *Far-end crosstalk (FEXT)* is the measure of signal interference at the transmit end of a twisted-pair cable from a transmit pair to an adjacent pair, as measured at the far end of the cable. *Power sum near-end crosstalk (PS NEXT)* is a measure of the coupling of undesired signal noise in a twisted-pair cable at the near end of multiple transmit pairs to an adjacent pair, as measured at the near end of the cable.

Tech Tip

Insertion loss is caused by conductors which touch but are not bonded together, as what happens when an RJ-45 patch cord is inserted into a patch panel. Attenuation is a continuous loss that occurs proportional to the length of a cable due to the (very low) resistance inherent in the cable. Both types of loss reduce the power level of a transmitted signal, and both are measured in decibels (dB). They are so similar that specifications and test equipment often use the two terms interchangeably.

Attenuation to crosstalk ratio far-end (ACR-F) is the ratio of signal strength to undesired signal noise in a twisted-pair cable, as measured at the far end of the cable. ACR-F is also known as equal level far-end crosstalk (ELFEXT). *Power sum attenuation to crosstalk ratio far-end (PS ACR-F)* is the ratio of signal strength to undesired signal noise in a twisted-pair cable at the near end of multiple transmit pairs to an adjacent pair, as measured at the far end of the cable. *Return loss* is the ratio of transmitted signal strength to signal strength reflected back to the transmitting end of a channel or permanent link. Return loss is measured in decibels (dB). **See Appendix.**

Summary

VDV technicians must know how to test copper VDV systems. There are three levels of testing. From least to most complex, the levels are verification, qualification, and certification. Verification and qualification tests are typically used for troubleshooting. Certification is used to confirm that a network is installed to a known standard. A structured cabling system must be certified in order to obtain a warranty from the OEM.

Chapter Review

1. What are the three categories of testing for copper VDV systems?

2. What are the two most common types of verification testers?

3. What types of conductors and cables can a toner be used with?

4. What types of cables can a wiremap tester be used with?

5. A wiremap tester consists of what two elements?

6. VDV technicians commonly use wire mappers to test for which five twisted-pair cable faults?

7. What is the purpose of a qualification tester?

8. Can a qualification tester be used to certify a network?

9. What types of cables can a certification tester be used with?

10. What are certification testers designed to measure and why?

11. When troubleshooting cables, what is the first test instrument a VDV technician typically uses after a visual inspection?

12. What is a toner?

13. What is the purpose of a toner probe?

Chapter Review

14. Describe a remote unit.

15. What is PoE?

16. Describe crossed pairs and split pairs.

17. What is a shorted pair?

18. What are the two most common types of coaxial cable faults?

19. What must be true of any network undergoing a qualification test?

20. List three common qualification tester diagnostic tests.

21. Most certification testers can test for what two standards?

22. What are two other names for certification testers?

23. A certification test is most commonly performed on what portion of a network?

24. Describe 11 typical certification test parameters.

25. What standards are most commonly used for certification tests of VDV systems in North America?

Chapter Activity Test Instrument Selection

Select the appropriate test instrument for each scenario. Possible test instruments are a toner, a certification tester, a wiremap tester, or a qualification tester. If more than one could be used, select the least complex test instrument.

Scenario 1: Which test instrument would a VDV technician use to locate a coaxial cable that is installed behind a wall?

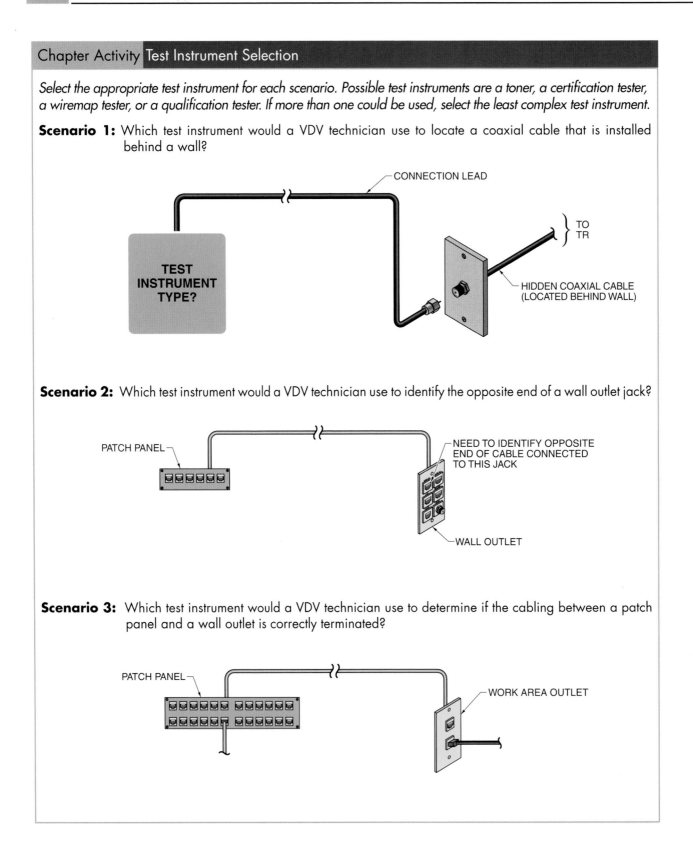

CONNECTION LEAD

TEST INSTRUMENT TYPE?

TO TR

HIDDEN COAXIAL CABLE (LOCATED BEHIND WALL)

Scenario 2: Which test instrument would a VDV technician use to identify the opposite end of a wall outlet jack?

PATCH PANEL

NEED TO IDENTIFY OPPOSITE END OF CABLE CONNECTED TO THIS JACK

WALL OUTLET

Scenario 3: Which test instrument would a VDV technician use to determine if the cabling between a patch panel and a wall outlet is correctly terminated?

PATCH PANEL

WORK AREA OUTLET

Chapter Activity Test Instrument Selection

Scenario 4: Which test instrument would a VDV technician use to determine if a cable has sufficient bandwidth to support a specific network?

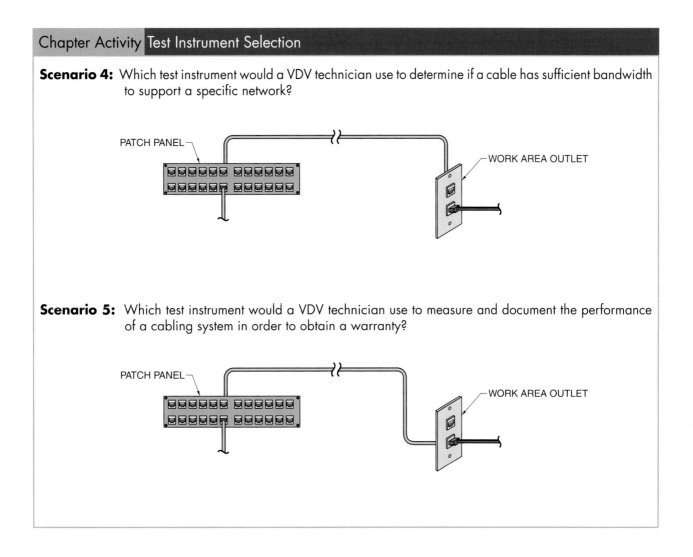

PATCH PANEL

WORK AREA OUTLET

Scenario 5: Which test instrument would a VDV technician use to measure and document the performance of a cabling system in order to obtain a warranty?

PATCH PANEL

WORK AREA OUTLET

Chapter 11

Installation and Termination of Fiber-Optic VDV Cabling Systems

Installing a fiber-optic system is similar to installing a copper system since the same cable-pulling techniques are used for both, the cables involved are similar in size. However, fiber-optic systems have different types of connectors, patching systems, terminations, and splicing methods. A VDV technician must be familiar with the equipment, material handling methods, and termination methods used to properly install fiber-optic systems.

OBJECTIVES

- Describe the general requirements for installing fiber-optic cable.
- Describe the differences between fusion splices and mechanical splices.
- Explain preterminated fiber-optic systems and how they apply to multifiber installations.
- Explain the different types of racking and patching equipment used with fiber-optic systems.
- Describe how to properly document all installations and related equipment.

Digital Resources

ATPeResources.com/QuickLinks
Access Code 838502

BACKBONE AND HORIZONTAL CABLING INSTALLATIONS

Fiber-optic cables are used in many different environments. They have a high data capacity and are lightweight. Also, because fiber-optic cables use light rather than electricity for signaling, they are not affected by electromagnetic interference (EMI). These properties make fiber-optic cabling a good alternative to copper cabling where weight limitations, capacity, and security from unauthorized access are a concern. Although fiber-optic cabling is most frequently used for outside plant (OSP) long-distance communications, it is used for indoor applications as well. VDV technicians will more commonly work with fiber-optic cables in indoor installations.

Indoor cables fall into two categories—backbone and horizontal. Fiber-optic cable is common in backbone cabling. Backbone cabling is used to connect telecommunications rooms (TRs), sometimes referred to as "closets." These TRs can be in the same building or in different buildings on the same campus, such as a hospital, commercial office complex, or school.

Fiber-optic cables are also used in horizontal cabling. Horizontal cables are those that are routed from a TR to individual work areas (computers). Although copper cabling, such as Cat 5e or 6A, is more commonly used for these types of installations, fiber-to-the-desk (FTTD) is also used. **See Figure 11-1.**

There are two main disadvantages to FTTD. The first is that the cost of the electronic equipment involved is higher since a fiber-optic network interface card (NIC) is needed in the desktop workstation. A *network interface card (NIC)* is a printed circuit board comprised of electronic circuitry used to connect a workstation to a local area network (LAN). Early versions of NICs consisted of a circuit board, or "card," that was added to a computer by either an external device or an internal expansion slot. Since almost all computers are now connected to the Internet, most have an NIC function built onto the motherboard. RJ-45 copper-based connections are almost universally used for this.

The second disadvantage with FTTD is that the power supply equipment required is different and more expensive than the equipment used with copper systems. Glass does not conduct electricity. The same property that makes a fiber-optic system resistant to EMI also prevents it from carrying electrical power to an end device. Therefore, separate power supplies must be used at both ends. The device at the end of the channel link must have a light source and conversion electronics to produce and interpret light signals.

The cost of cables and connectors has decreased over time and is now comparable to copper UTP equipment. In some cases, fiber-optic cables and prepolished connectors are easier to install than UTP equipment. However, in addition to NICs, fiber-optic systems also require transceivers, Ethernet hubs, switches, routers, and patching equipment (short lengths of cable with fiber-optic connectors used for connections to equipment racks). This equipment is significantly more expensive than comparable equipment used with copper systems.

Due to the increase in the use of Power over Ethernet (PoE) devices such as security cameras and Voice over IP (VoIP) telephones, copper cables remain the most logical choice for horizontal cabling installations. However, for some applications, such as those that require an extremely high degree of communication security or a high bandwidth, fiber-optic cables are preferred. With either type of installation, backbone or horizontal, fiber-optic cables are installed and routed with the same basic procedures as copper cables.

Fiber-optic network interface cards are a requirement for connecting desktop workstations to the Internet.

Fiber-to-the-Desk (FTTD)

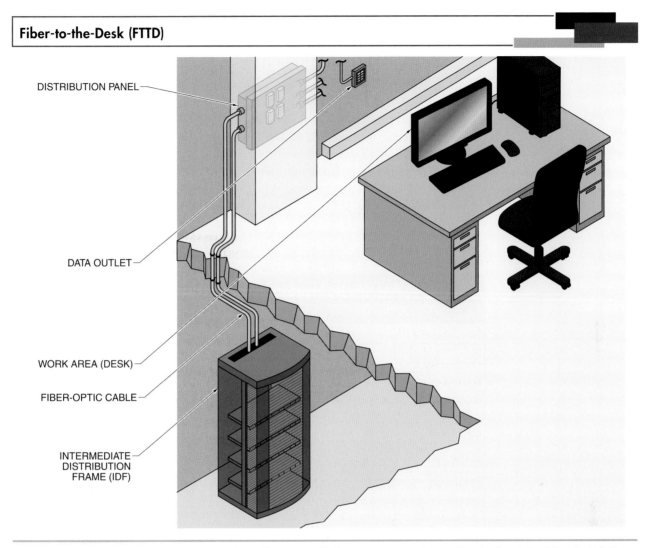

DISTRIBUTION PANEL

DATA OUTLET

WORK AREA (DESK)

FIBER-OPTIC CABLE

INTERMEDIATE DISTRIBUTION FRAME (IDF)

Figure 11-1. Although copper cabling, such as Cat 5e or 6A, is more commonly used for horizontal cabling installations, fiber-to-the-desk (FTTD) is also used.

FIBER-OPTIC CABLE DESIGN AND INSTALLATION

A fiber-optic strand is composed of a core, a cladding, and a coating. A fiber-optic cable is composed of a single strand or multiple strands along with a strength member and a jacket. Usually, however, they are constructed with multiples of six strands. Groups of strands are covered with a buffer tube. The buffer tube separates a group of strands from other groups for protection and identification purposes. The cable has a strength member in the center to provide stability and a way to attach a pulling mechanism for installation purposes. The jacket is for protection and identification of the cable.

Fiber-optic cables usually contain aramid yarn. Aramid yarn is a strength member that is heat resistant and has good tensile strength. It also provides a method for pulling the cable into place without damaging the relatively fragile glass strands.

Tech Tip

Aramid yarn, also known by the trade name Kevlar®, was originally developed by the DuPont Chemical Company for military applications. It is the primary strength component in bulletproof clothing used by military and law-enforcement personnel.

Though most VDV fiber-optic cables contain these parts, the actual design can vary with different types of cable and different cable OEMs. All OEMs, however, adhere to a strand color standard referenced in TIA-606-A so strands can be identified easily. **See Figure 11-2.** If there are 6 strands in each buffer tube, then the first 6 colors are used. If there are 12 strands, then all 12 colors are used. In addition, buffer tubes are often colored using the same convention as the individual strands to differentiate one cable from the next.

Fiber-Optic Strand Color Standard	
Fiber Number	Color
1	Blue
2	Orange
3	Green
4	Brown
5	Slate
6	White
7	Red
8	Black
9	Yellow
10	Violet
11	Rose
12	Aqua

Figure 11-2. Fiber-optic cable OEMs adhere to a strand color standard so strands can be identified easily.

Parallel Optics

With fiber-optic cables, the fibers are sometimes laid out in parallel, which involves parallel optics. *Parallel optics* is a fiber-optic technology design that uses multiple fibers for a single data stream. Parallel-optical design uses 12 strands of fiber next to one another in a harness which allows the separate strands to be branched out for connection to a module. This type of design is used in preterminated fiber systems (plug-and-play) and in field-terminated,

multifiber connectors, such as duplex or MTP connectors.

Plug-and-play modules provide a quick-connecting interface with plastic body cassettes, which are used for collecting individual fibers into a multifiber cable, between multiple fiber-optic connectors on a communications path (trunk) and the LC duplex, SC duplex, or mutiple fiber push-on (MTP) jumper cables. These then connect directly to the electronic devices. **See Figure 11-3.**

When installing a plug-and-play module, the VDV technician routes the fiber-optic cabling in place and then only needs to plug it into the module to make it operational. Often, testing is still performed to confirm that the link operates as designed, but no field termination of connectors is necessary.

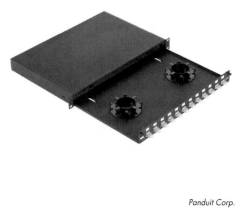

Panduit Corp.

Preloaded connector housings are used with parallel optics.

Plug-and-Play Systems

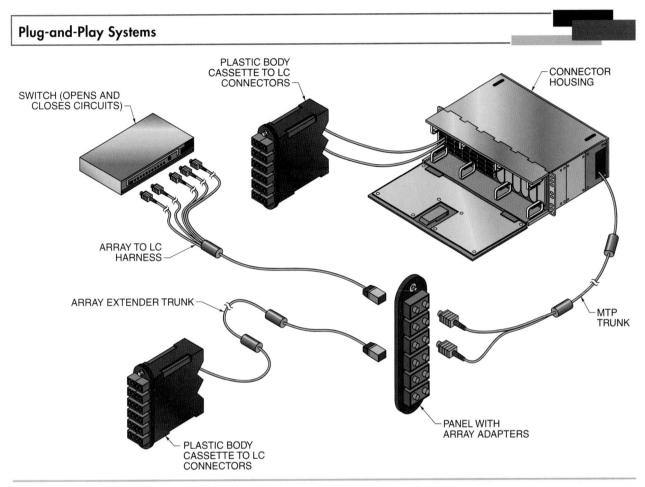

Figure 11-3. Plug-and-play modules provide a quick-connecting interface with plastic body cassettes between multiple fiber-optic connectors on a communications path (trunk) and the LC duplex, SC duplex, or MTP jumper cables. These then connect directly to the electronic devices.

Speed of Data Transmission

The speed of data transmission is affected by bandwidth. Due to modal dispersion, there is a practical speed limit on fiber-optic networks of about 10 Gbps (10,000,000,000 bits per second). This is a dramatic improvement over the original speed of a 2400 bps modem of only a few decades ago. However, it is anticipated that speeds of 40 Gbps and 100 Gbps will soon be utilized in data centers. A *data center* is a building or part of a building that is designed, built, and maintained specifically to house and connect computer servers. Data centers have become essential to the technology now being used in everything from entertainment and gaming to science and research, as well as to the containment of classified information and transmission.

To reach these improved speeds, parallel-optical technology is required. One glass fiber does not have the needed data-carrying capacity, so multiple fibers connected in parallel are used to achieve the required bandwidth. For example, if a speed of 40 Gbps is required, then 4 fibers per direction of data movement (4×10 Gbps $= 40$ Gbps) are needed. If a speed of 100 Gbps is required, then 10 fibers are needed. Because data flows in two directions, a total of 8 fibers or 20 fibers respectively would be required for these specific applications.

In order to connect multiple fibers concurrently, OEMs produce connectors that have 12 or 24 strands. These connectors allow an immediate, simultaneous connection with parallel optics. The reason for 12 or 24 strands (rather than 8 or 20) is that fiber-optic cable is manufactured with multiples of 6 strands. *Note*: For parallel-optical applications, the extra fibers would not be used.

Currently, data centers do not have the technology to effectively use these speeds. The advantage of using these connectors is that plug-and-play hardware can be used to distribute the fibers into standard duplex (two-fiber) patch panels, which are easy to install and provide an inexpensive option for future upgrades. Almost all fiber-optic equipment OEMs now provide plug-and-play systems for specific cable lengths. In addition, most OEMs provide connectors that can be field terminated.

Pulling Methods and Cable Support Systems

Fiber-optic cable is pulled into place in the same manner as copper cable. The aramid yarn in fiber-optic cable provides good pulling strength. Fiber-optic cable support systems include open cable trays, closed fiber ducts, J-hooks, slings, and cables integrated with a steel messenger cable or wire. **See Figure 11-4.**

Cables integrated with steel messenger cables or wires are typically used for outdoor, aerial applications, where a sturdy cable that has extra support is required. If an installation is performed through an open cable tray or with J-hooks, an installer can simply grasp the cable and pull it into place. This is known as hand pulling. Most fiber-optic cable installation guidelines have a suggested pull force of no more than 25 lb.

If a cable is installed in an enclosed support system such as EMT, a pulling string or rope must be used. The method of connecting the pull string or rope to the cable is important. The outside jacket of the cable is a durable protection for the more fragile glass strands inside, but the strands are not physically connected to it. When pulling through EMT, an installer will most likely exceed the hand-pulling tension limit of 25 lb. If the pull string is only attached to the outer jacket, it may stretch or tear the cable.

To prevent damage to the strands, the pull string must be attached to the aramid yarn by separating the yarn fibers, knotting them to a pulling eye, and then taping the outer jacket with electrical tape. **See Figure 11-5.**

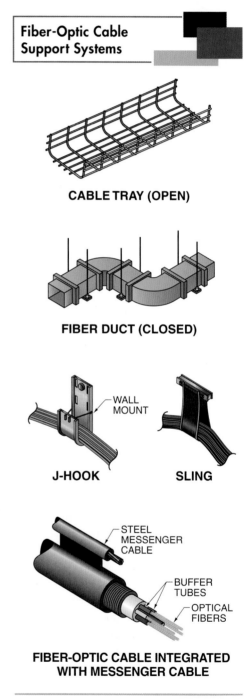

Fiber-Optic Cable Support Systems

CABLE TRAY (OPEN)

FIBER DUCT (CLOSED)

WALL MOUNT

J-HOOK   **SLING**

STEEL MESSENGER CABLE

BUFFER TUBES

OPTICAL FIBERS

FIBER-OPTIC CABLE INTEGRATED WITH MESSENGER CABLE

Figure 11-4. Fiber-optic cable support systems include open cable trays, closed fiber ducts, J-hooks, slings, and cables integrated with steel messenger cables or wires.

Pulling Fiber-Optic Cables

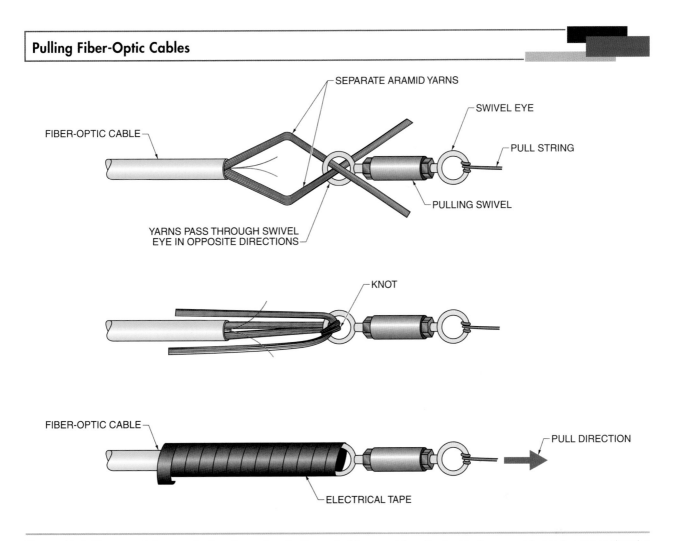

Figure 11-5. To prevent damage to the glass strands, the pull string must be attached to the aramid yarn by separating the yarn fibers, knotting them to a pulling eye, and then taping the outer sheath with electrical tape.

The length of cable that can be pulled depends on many factors in an application, such as the type of cable (indoor, OSP, ribbon, etc.), the raceway, the environment temperature, and how straight the run is. These factors all affect the coefficient of friction. With the exception of short pulls, cables should be lubricated to reduce friction in order to decrease pulling tension. Lubricants should be approved by the cable manufacturer—lubricants for copper communication or electrical cables may damage fiber-optic cables. Some lubricant manufacturers offer online calculators to help VDV technicians decide which lubricant is the best to use for a particular application.

Most cables can only be safely pulled with a swivel pulling eye attached to the aramid yarns. Swivels are important to reduce twisting loads on a cable. For high-tension loads, a breakaway swivel is used to prevent damage if pulling tension exceeds the maximum tension allowed by the cable specification. Some cables are rated for pulling by the jacket if a flexible wire mesh grip is used. These are also usually attached to the aramid yarns of the cable.

Tension for pulling outdoor cables is often high enough that a powered drum with a pulley can be used for all but the shortest pulls. The drum should be equipped with a monitor that can shut down the mechanism if a preset tension is exceeded. If the tension becomes too high, the cause of the excess friction must be corrected.

During installation, the most common cause of damage to a fiber-optic cable comes from bending a cable past its minimum bend radius. With proper care and handling, fiber-optic cable can be reliably installed to avoid this. Installation pulleys and gentle sweeping turns in cable trays and conduit systems are typically used. Although fiber-optic cable is not overly fragile, to avoid permanent damage, it cannot be bent past its minimum bend radius. **See Figure 11-6.**

Fiber can also be installed through the use of blown fiber. *Blown fiber* is an installation method in which a housing containing many smaller tubes is installed without fiber. An air compressor is then used to move the fiber into and through the tubes. Using compressed air reduces the amount of damage to the fiber since no pulling on the fibers or cable is required. This method also reduces the need for splices and allows for additions to the installation to be made at a later time. Blown fiber is typically used for underground installations.

TERMINATIONS

In order to connect fiber-optic cables and equipment, the cables must be terminated. A termination places a connector on the end of a strand or strands of glass so that the strands can be aligned with others, or with a light source or the light-detection electronics involved. When terminated with matching connectors, the cables can be connected and disconnected repeatedly and reliably.

Connector Types

The most commonly used fiber-optic connectors are single-fiber connectors used in pairs, known as duplex connectors (which transmit from one end of a cable and receive on the opposite end). Duplex LC connectors are currently used more often than other connector types. They have a small form factor (or smaller connector body), which allows a greater number of them to be installed in a small patch panel. A high density arrangement of connectors increases available rackspace.

Minimum Bend Radius

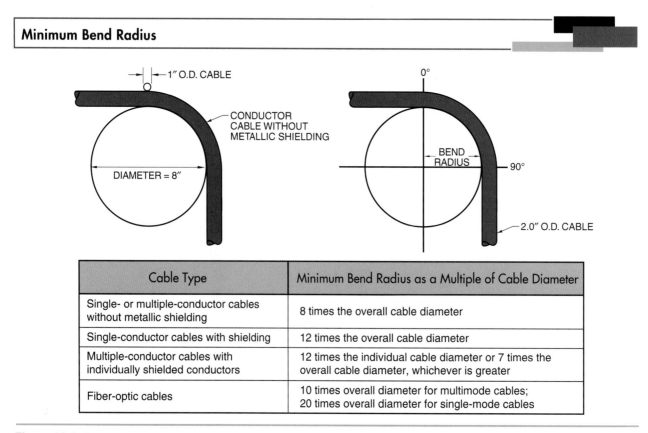

Cable Type	Minimum Bend Radius as a Multiple of Cable Diameter
Single- or multiple-conductor cables without metallic shielding	8 times the overall cable diameter
Single-conductor cables with shielding	12 times the overall cable diameter
Multiple-conductor cables with individually shielded conductors	12 times the individual cable diameter or 7 times the overall cable diameter, whichever is greater
Fiber-optic cables	10 times overall diameter for multimode cables; 20 times overall diameter for single-mode cables

Figure 11-6. Failure to maintain a minimum bend radius can cause signal loss or glass breakage within a fiber-optic cable.

Although all fiber-optic connectors work on the same fiber-optic strands, they are not intercompatible. For example, an LC connector will only connect to another LC connector. All connector OEMs also produce termination kits that are specific to their connectors. However, a terminated LC from one OEM will connect to an LC from another OEM.

As previously described, connectors are attached to fibers in a few different ways. Field-polished connectors are rare. They require a great amount of time to polish and install. Mechanical connectors, fusion-spliced pigtails (with factory-installed connectors), and fuse connectors are common. The most prevalent are mechanical connectors. These have a lower signal loss, provide a factory-polished end, and are easy to install them making them a low-cost option.

Multifiber Connectors

Multifiber push-on (MPO) connectors are connectors that are compliant with standard IEC-61754-7. They usually house 12 or 24 fibers. Though the standard allows for the connection of up to 72 strands, a 72-strand connection is rarely used.

Recently, fiber-optic equipment OEMs have introduced new connectors and patching hardware for use in data centers. Data center design levels are referred to as tiers. A Tier 1 data center is the simplest and least expensive tier. It has connection points and processing equipment, and it has an annual downtime of about 29 hr (99.671% availability).

A Tier 2 data center has everything a Tier 1 data center has, but it also helps control a building's heating and cooling systems. It has an annual downtime of about 23 hr (99.741% availability). A Tier 3 data center has all the capabilities of a Tier 2 data center plus an added electrical path, thereby providing additional infrastructure. Tier 3 data centers have an annual downtime of about 1½ hr (99.982% availability). A Tier 4 is the most complex and costly type of tier. In a Tier 4 data center, all the components are fully fault-tolerant including uplinks, storage, chillers, HVAC systems, and servers. Everything is dual-powered, and the annual downtime is typically around 30 min (99.995% availability). **See Figure 11-7.**

Data Center Tiers

Small Commercial Business
Tier 1 — Basic
99.671% availability
Susceptible to disruptions
Single path for power
No redundant components

Medium Commercial Business
Tier 2 — Redundant
99.741% availability
Less susceptible to disruptions
Single path for power
Redundant components

Large Commercial Business
Tier 3 — Additional Redundant
99.982% availability
Planned activity without disruption
Dual-powered
Redundant components

Multi-Million Dollar Corporation
Tier 4 — Fault-Tolerant
99.995% availability
Can withstand at least one worst-case event
Dual-powered
Redundant components

Figure 11-7. Data center design levels are referred to as tiers. A Tier 1 data center is the simplest type of tier, while a Tier 4 is the most complex.

MTP® connectors were developed specifically for applications such as data centers. An MTP connector is a fiber-optic connector that has the same size as an SC connector but provides 12 times the density, making it ideal for high fiber-count data center and parallel-optical connections between servers. MTP connectors are a variation of MPO connectors and are used specifically for indoor cabling and device connections. **See Figure 11-8.**

MTP Connectors

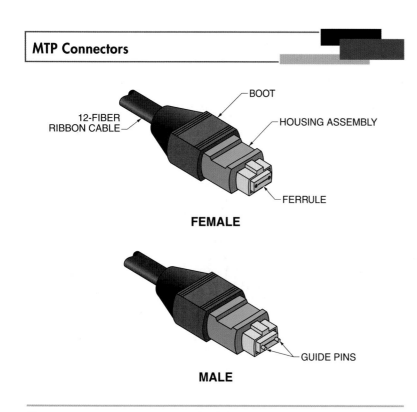

12-FIBER RIBBON CABLE

BOOT

HOUSING ASSEMBLY

FERRULE

FEMALE

GUIDE PINS

MALE

Figure 11-8. An MTP connector is a fiber-optic connector that has the same size as an SC connector but provides 12 times the density, making it ideal for high fiber count data center and parallel-optical connections between servers.

Redundant capacity components in data centers include cabling infrastructure, transmission electronics, and switching components. All of these components have sufficient capacity to fully default to redundant (currently unused) identical components should a failure occur in the primary (currently used) components.

Tier 1 data centers typically have no redundant components. Redundant components are present in Tier 2, Tier 3, and Tier 4 data centers.

MTP connectors are designed for improved optical signal performance over MPO connectors. MTP connectors have a free-floating ferrule, which helps maintain physical contact and reduces signal loss. The MTP connector ferrule can float in the MTP to maintain contact on a mated pair if there is a strain on the cable. MPO connectors do not have this feature.

MTP connectors are completely compliant with all MPO connector standards including EIA/TIA-604-5 FOCIS 5 and IEC-61754-7. They are interchangeable with all generic MPO connectors that are compliant to these same standards.

SPLICING

Splicing optical fibers is a method of making a permanent connection between two separate fibers so that the light from one can be transmitted to the other. Splices are also performed to repair damaged cables, lengthen cables, or permanently attach a short length of cable to a device such as a connector. Fiber-optic splices can be fusion splices or mechanical splices. The quality of a splice is affected by the differences in the qualities of the glass in each cable (intrinsic factors) and the alignment and mating of the fibers (extrinsic factors). All splices are considered permanent connections.

Note: To prepare for a splice procedure, the work area must be clean, dry, and adequately lit. Work must be performed over the fiber-optic mat included in the termination or tool kit, and any scrap must be disposed of in the proper fiber-optic disposal container. Safety glasses must always be worn to prevent glass fibers from entering the eyes.

Fusion Splicing

A fusion splice is a fiber-optic splice formed by applying heat to the ends of two different fibers in order to form one continuous strand. The strands of fiber are cleaved, chemically cleaned, aligned, and then electrically melted together with a fusion splicer. A *fusion splicer* is a device that uses a high-voltage electrical arc between two electrodes to heat and melt the ends of glass fibers in order to join them and form a continuous strand. A fusion splicer includes the electrodes as well as an alignment and splicing tool. **See Figure 11-9.** This produces an extremely low signal loss connection where the fibers are joined. Splicing optical glass fibers with a fusion splicer is performed by applying the following procedure:

1. Turn the fusion splicer on, and select the profile for the fiber to be spliced.

2. Slide the protective heat-shrink tube or boot over one end of the fiber, and slide the boot away from the area to be spliced.

3. Use a fiber stripper to strip about 1″ of buffer and coating off the glass fiber.

4. Clean the stripped fiber end using a lint-free cloth soaked with an optical-fiber cleaning solution.

5. Place the cleaned fiber in a fiber-cleaving unit, as specified by the cleaving unit OEM instructions.

6. Cleave the fiber.

7. Place the cleaved fiber in the fusion splicer as specified by the fusion splicer OEM.

8. Repeat steps 2–7 with the other fiber to be spliced.

9. Close the cover and perform the fusion splice as specified by the fusion splicer OEM.

10. Move the heat-shrink tube or boot over the spliced area, and place the fiber in a heat-shrink oven to seal and protect the splice from damage.

11. Turn the fusion splicer off.

12. Remove the fiber from the heat-shrink oven.

Tech Tip

Fusion splices must never be made in an enclosed space, such as in a manhole, or in an explosive atmosphere. Fusion equipment is too large for most aerial applications, so fusion splices are usually made above ground in a work area set up for this purpose.

Fusion Splicers

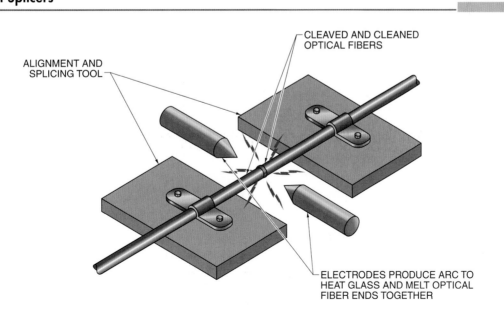

ALIGNMENT AND SPLICING TOOL

CLEAVED AND CLEANED OPTICAL FIBERS

ELECTRODES PRODUCE ARC TO HEAT GLASS AND MELT OPTICAL FIBER ENDS TOGETHER

Figure 11-9. A fusion splicer creates a fiber-optic splice by applying heat to the ends of two different fibers in order to form one continuous strand.

A finished splice must then be protected from damage. Usually, a splice tray is used, where up to 12 strands can be encased in a rigid plastic housing for protection. A *splice tray* is a device used to organize and protect spliced fibers. **See Figure 11-10.** If the splices are to be used for indoor applications, the housing can be mounted in a rack or in a wall. If the splices are to be used for outdoor applications, an environmentally safe housing is used. The main disadvantage of fusion splicing is the high cost of the fusion splicer.

Splice Trays

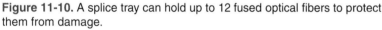

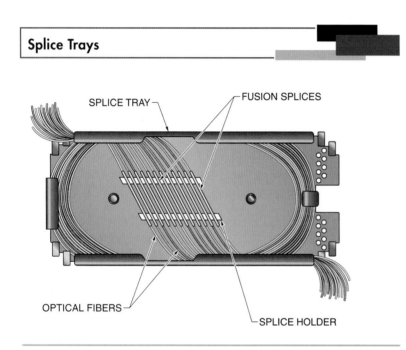

Figure 11-10. A splice tray can hold up to 12 fused optical fibers to protect them from damage.

Mechanical Splicing

A *mechanical splice* is a fiber-optic splice in which the ends of two different fibers are mechanically connected. **See Figure 11-11.** Mechanical splices are made by gluing or crimping the connectors that hold the strands together. The fiber ends are placed in a splice assembly component and then into an assembly tool to make the splice. The splice assembly component contains an index-matching gel needed to complete the process.

A mechanical splice is similar to a fusion splice but does not use heat to join the two strands. Instead, they are mechanically aligned and held in place. A mechanical splice creates a higher degree of signal loss (typically up to 1 dB) than a fusion splice, but it is adequate for most applications. Mechanical splices are often used due to costs; the equipment required to make mechanical splices is considerably less expensive than the equipment required to make fusion splices. A mechanical splice with optical glass fibers is performed by applying the following procedure:

1. Place the splice assembly component into the assembly tool per the OEM's instructions.
2. Use a fiber stripper to strip about 1″ to 1⅛″ of coating and buffer from the glass fiber.
3. Clean the stripped fiber ends with an optical-fiber cleaning solution and a lint-free wipe.
4. Place the cleaned ends, one at a time, in the fiber cleaver and cleave.
5. Push one end of a cleaved fiber into one end of the splice assembly component and gently push until no further movement is felt.
6. Repeat Step 5 with the other cleaved fiber end.
7. Place both fibers and the splice assembly component in the assembly tool.
8. Pull the assembly tool handle down and apply pressure until the splice is complete.
9. Gently remove the spliced fiber assembly from the assembly tool.

PRETERMINATED FIBER SYSTEMS

Preterminated fiber systems (plug-and-play systems) require almost no tools and minimal field expertise to install. The cable is typically ordered in specific lengths with preterminated multifiber connectors. The cables are shipped with protective covers over the preterminated ends to protect them during transportation and installation. The installer simply pulls the cable from one point to another and plugs it into a compatible cassette.

Mechanical Splices

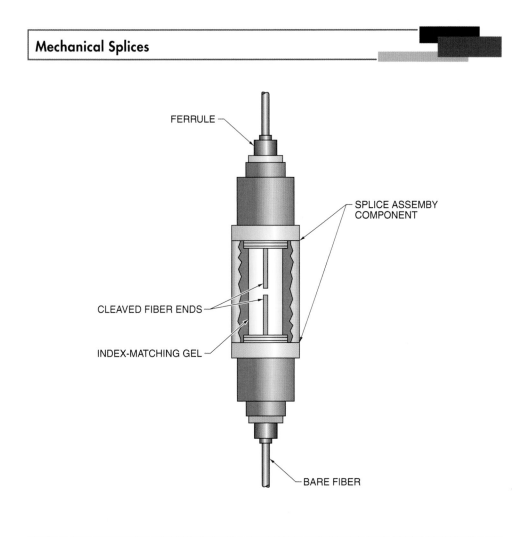

FERRULE

SPLICE ASSEMBY COMPONENT

CLEAVED FIBER ENDS

INDEX-MATCHING GEL

BARE FIBER

Figure 11-11. With a mechanical splice, the ends of two glass fibers are mechanically connected.

The cable is more expensive than other types, but this added cost is usually more than compensated by the ease and speed of installation. There are however a few important details to be aware of, such as the cassettes, polarity, and preterminated cable lengths.

Cassettes

With preterminated fiber-optic systems, multifiber cables with multiple single-fiber connectors are plugged into cassettes at the large end. The cassettes are composed of a rigid plastic housing. Each has a single MTP connector on the back panel and single-fiber connectors on the front panel. Inside a cassette, the fibers are split into pairs, routed, and preterminated to the single-fiber connectors.

The cassette provides a patching point for connecting individual pairs to transmission equipment. **See Figure 11-12.**

Polarity

In a fiber-optic network, polarity refers to the directions of transmission paths between the transmitters and receivers of channels. Due to the use of fiber in pairs, the manner in which the connectors are aligned can affect performance. In a 12-strand cable, all of the fibers are fixed in place. The first fiber is in position 1 on both ends, the second fiber is in position 2 on both ends, and so forth. When the fibers are connected to equipment, the transmitter and receiver are also fixed in place.

Cassettes

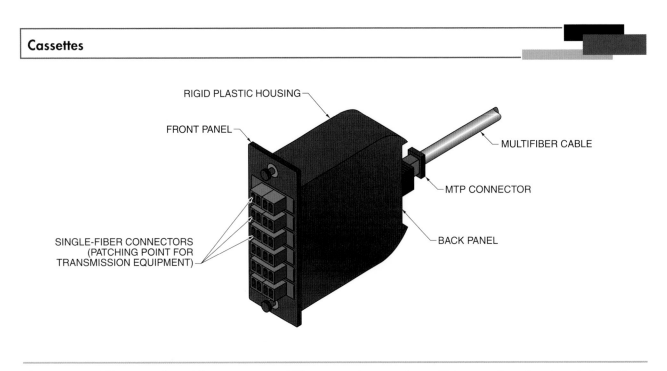

RIGID PLASTIC HOUSING

FRONT PANEL

MULTIFIBER CABLE

MTP CONNECTOR

SINGLE-FIBER CONNECTORS
(PATCHING POINT FOR
TRANSMISSION EQUIPMENT)

BACK PANEL

Figure 11-12. With preterminated fiber systems, multifiber connectors are plugged into cassettes at each end of a cable.

Depending on the cassette type, every pair of fibers must be cross-connected to ensure that transmitters are connected to receivers at the other end. Thus, a straight cassette is used on one end, and a reversing cassette is used on the other. A straight cassette passes fiber-optic strands through in the same order as the incoming cable. A reversing cassette swaps the position of every other fiber. In this way, a transmitter on one end is always connected to a receiver on the other. **See Figure 11-13.**

Preterminated Cable Lengths

While not as critical as cassette design or polarity, the lengths of the cables installed must be considered. If a cable for a specific application is too short, modifications to the cable or a replacement will be required. If a cable for a specific application is too long, the excess must be stored (coiled), which creates a storage issue. Also, because fiber-optic cable loses signal strength proportionate to its length, the extra cable introduces unnecessary signal loss.

> **Tech Tip**
>
> In a fiber-optic network, polarity is typically tested by de-energizing the equipment and then using a continuity tester to check each fiber.

CABLE RACKS

In order to organize and efficiently utilize fiber-optic systems, racking and patching hardware are used. Cable racks are hardware devices used to organize the cables in a network system. Cable racks and hardware used for fiber optics are similar to those used with copper systems, but they are designed for use with smaller, lighter fiber-optic cables.

Cable racks range in height from 39″ to 84″ and in width from 19″ to 23″, with 19″ being the most common. Mounting holes in the racks are typically spaced ⅝″ to 2″ apart, which allows for the installation of equipment from different OEMs. Racks are available in wall-mount designs, open-frame designs, or as complete cabinets.

Polarity

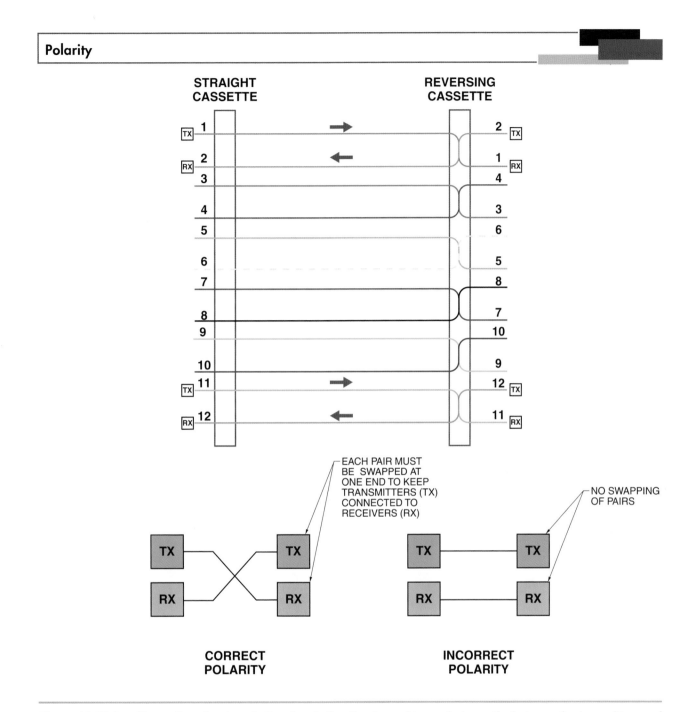

Figure 11-13. In a fiber-optic network, polarity refers to the directions of transmission paths between the transmitters and receivers of channels.

FIBER-OPTIC CABLE MANAGEMENT

Proper cable management is more critical with fiber-optic cable than with copper cable because fiber-optic cable is more susceptible to damage. It can be damaged if crushed or if the minimum bend radius is not maintained. This damage can cause signal degradation or cable failure, which leads to service interruption or downtime. When commercial systems are involved, the cost of repairing a damaged cable must be added to the cost of lost business.

Many OEMs have developed integrated cable-management or cable-routing systems specifically for fiber-optic cables. These systems consist of a channel or tray with covers (typically hinged), right angle (90°) fittings, 45° fittings, tee fittings, cross fittings, horizontal-to-vertical transition fittings, and mounting hardware. The routing systems are typically composed of a rigid plastic material. These systems are available in various sizes to accommodate different quantities of cable. The routing systems can be mounted to struts, to a ladder rack, or to the top of an equipment cabinet. **See Figure 11-14.**

There are many advantages to using a fiber-optic cable-management system. First, it provides physical protection for fiber-optic cables. Second, all of the components of a fiber-optic cable-management system are designed to maintain the minimum bend radius of the cable. This prevents bends that are too tight which may cause damage. Third, the system provides a clear, well-defined path between equipment, cabinets, and racks. This reduces congestion in the horizontal and vertical channel or tray. Also, the cables are less likely to become tangled. Finally, the system provides an easy access to the fiber-optic cables for troubleshooting, moves, additions, and changes.

Fiber-Optic Cable Management Systems

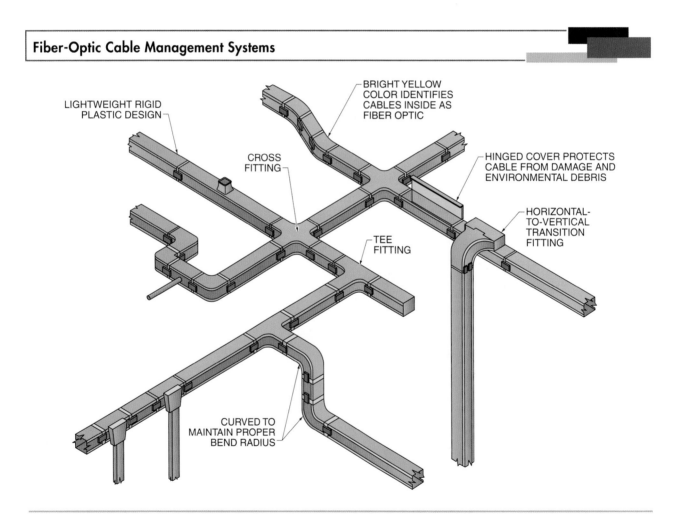

Figure 11-14. Integrated cable-management or cable-routing systems developed specifically for fiber-optic cables consist of a channel or tray with hinged covers, angled fittings, tee fittings, cross fittings, horizontal-to-vertical transition fittings, and mounting hardware.

Housing Enclosures

A housing enclosure provides a connection point at the end of a fiber-optic cable and can accommodate multiple connectors. It is where one cable can be connected to another or to a piece of equipment, and it provides physical protection for the fiber ends. (As the glass ends are extremely small, they are subject to dust and physical harm.) Most housing enclosures have attachments for cable jackets in order to provide strain relief, storage spaces for excess fiber needed for termination or splicing, and a way to mount the connector panels to align the cores of the fibers to the cores of the patch cords. Connectors are mounted to a removable shelf or tray within the enclosure.

Housing enclosures can be wall or rack mounted, depending on the facility. In large data centers, rack-mounted units are most commonly used. This allows the cable to be routed through the cable management system and placed as close as possible to the end equipment. **See Figure 11-15.** In smaller TRs, such as in residential or light commercial installations, there is a smaller quantity of equipment and fiber. In these cases, wall-mounted units are often used.

Patch Panels and Distribution Panels

A *patch panel* is a device used to make connections between incoming and outgoing VDV lines. A patch panel provides a way to connect horizontal or backbone cables to a group of fixed connectors, which may be accessed by cables to form cross-connections or interconnections. Patch panels allow connections to be made or broken at the desire of the network administrator. A patch cord is used to make the connection between a connector in the patch panel and the cable that is routed to the device.

Patch panels are installed within housing enclosures, and they are ordered for a specific type of connector. For example, patch panels with LC connectivity would be required for systems that use LC connectors and jacks. Patch panels are arranged in pairs for two-way communication. A typical patch panel provides the termination and patching point for 12 strands of fiber (6 duplex connections). Housings are sized by the number of patch panels they can accommodate.

Housing Enclosures

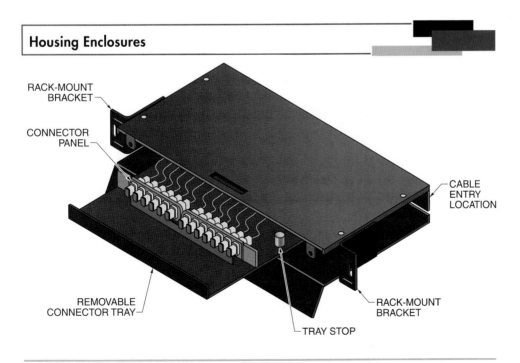

RACK-MOUNT BRACKET

CONNECTOR PANEL

CABLE ENTRY LOCATION

REMOVABLE CONNECTOR TRAY

TRAY STOP

RACK-MOUNT BRACKET

Figure 11-15. A housing enclosure provides a connection point at the end of a fiber-optic cable and can accommodate multiple connectors.

A distribution panel is similar to a patch panel, but the connections involved are treated as permanent connections. Although the connections in a distribution panel can be changed, the decision to do so must be authorized by someone in authority. A distribution panel typically has a locking mechanism on the front to prevent unauthorized access. **See Figure 11-16.**

Tech Tip

Fiber-optic patch-and-splice enclosures have many different names including termination panels, patch-and-splice panels, fiber-splice boxes, splice distribution boxes, fiber-splice panels, patch panels, fiber-optic panels, patch-and-splice modules, fiber-splice closures, and fiber-splice chassis.

Patch Panels and Distribution Panels

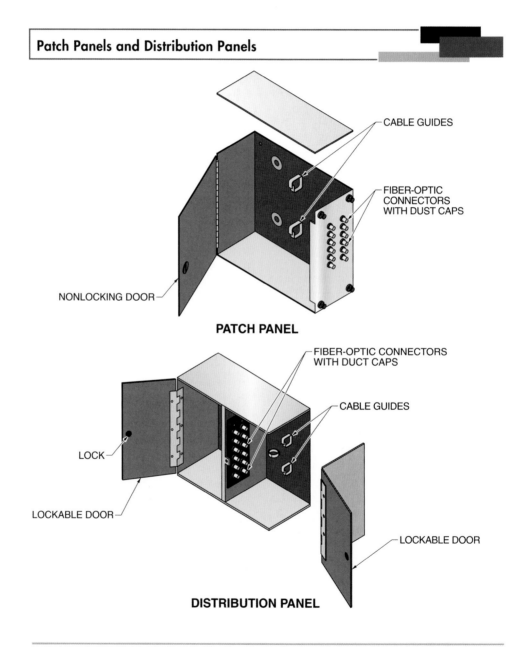

CABLE GUIDES

FIBER-OPTIC CONNECTORS WITH DUST CAPS

NONLOCKING DOOR

PATCH PANEL

FIBER-OPTIC CONNECTORS WITH DUCT CAPS

CABLE GUIDES

LOCK

LOCKABLE DOOR

LOCKABLE DOOR

DISTRIBUTION PANEL

Figure 11-16. Patch panels and distribution panels are connection hardware devices that provide a means to connect horizontal or backbone cables to a group of fixed connectors, which may be accessed by cables to form cross-connections or interconnections.

Patch Cords

As with copper systems, fiber-optic systems use patch cords to connect backbone and horizontal cables to patch panels or distribution panels. A *patch cord* is a flexible 3′ to 12′ length of cable used to connect a network device to a main cable run or a panel. In fiber-optic systems, patch cords are two-strand fiber-optic cables and are available from OEMs in standard or custom lengths. The cables require no tools to install or manage. The physical locking mechanism between the connector and the jack holds the connector in place. Detailed housekeeping, records of fiber-optic paths, and use of cable labeling systems are essential to tracking and making changes to interconnected fiber-optic systems.

DOCUMENTATION

The larger the fiber-optic installed base is, the more important it becomes to document all installation details. In large data center applications, the ability to add additional equipment may be limited by the volume and endpoints of fiber-optic pathways.

There are several software products available to help track fiber-optic pathways. Usually, however, a carefully maintained spreadsheet is adequate. Each cable is listed with a description of the endpoints, the length, the fiber count, the type of fiber, and the type of connectors. The used fibers are noted in a column with the connection date and what they are connected to. Dark fibers are also noted. Cross referencing one cable record to another provides a full path description as well as any unused capacity. A VDV technician will usually help establish the original documentation, which is then maintained by the owner or a representative.

FIBER-OPTIC SYSTEM INSTALLATION AND SAFETY CONSIDERATIONS

When installing fiber-optic systems and equipment, technicians must take a number of things into account. The optical ratings of the equipment involved must match,

the installed cable must be appropriate for the application environment, there must be protection from unterminated fibers, and the job site must be safe.

VDV technicians must always verify the ratings of the fiber and equipment being installed as single-mode or multimode so they are not mismatched. They must also verify that indoor-rated cable is installed for indoor applications only.

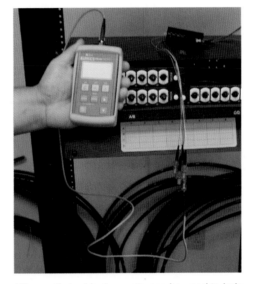

Fiber-optic test instruments can be used to help document the installation details of individual fiber strands.

Unterminated fibers can accidentally get in the eyes. VDV technicians must always wear protective eyewear and other appropriate PPE to prevent difficult-to-see glass fibers from entering the eyes. The ends of unterminated fibers must be covered to prevent glass breakage or visual exposure to light beams. Specific job site safety precautions must always be followed. Specific job site safety precautions may involve electrical equipment, ladders, trenches, and chemical exposure.

In addition to these safety considerations, the Fiber Optic Association also recommends the following:

- Keep all food and beverages out of the work area. If fiber particles are ingested, they can cause internal hemorrhaging.
- Work on a black work surface to make fiber scraps easier to see.
- Wear disposable aprons to minimize fiber particles on clothing. Fiber particles on clothing can later get into food, drinks, or be ingested by other means.
- Never look directly into the end of a fiber cable until verifying that there is no light source at the opposite end. Use a fiber-optic power meter to make certain the fiber is dark. When using an optical tracer or continuity checker, look at the fiber from an angle at least 6″ away from the eyes to determine if the visible light is present.

- Only work in well-ventilated areas.
- Do not handle contact their lenses until hands have been thoroughly washed.
- Do not touch the eyes while working with fiber-optic systems until hands have been thoroughly washed.
- Keep all combustible materials safely away from curing ovens.
- Always handle cut fiber pieces with tweezers.
- Put all cut fiber pieces in a properly marked container for disposal.
- Thoroughly clean the work area when finished.
- Do not smoke while working with fiber-optic systems.

Summary

Fiber optics is a common technology used in communications. Light pulses from a transmitter are conveyed to and detected by a receiver. Connectors, cables, and cords must be properly installed and terminated in order for a system to operate properly. Between the transmitter and receiver, the path must be continuous with a low amount of signal loss. The number of connection points, splices, and patch connections only matter to the extent that they introduce signal loss.

New data centers are almost exclusively using fiber optics. The management of a fiber-optic system has become an extremely important part of a data center manager's job. The proper design and installation of a fiber-optic system by a VDV technician is what makes that possible.

Chapter Review

1. What are the main properties of fiber-optic cables?

2. What are the two categories of indoor cables?

3. What are the two main disadvantages of FTTD?

4. In addition to NICs, what other pieces of equipment are required for fiber-optic systems rather than copper systems?

Chapter Review

5. In addition to providing strength, what is the purpose of the aramid yarn in a fiber-optic cable?

6. What fiber-optic technology design uses multiple fibers for a single data stream?

7. Due to modal dispersion, what is the practical speed limit of a fiber-optic network using parallel-optic technology?

8. How does parallel-optical technology increase the speed of a fiber-optic network?

9. List five types of fiber-optic cable support systems.

10. What is the most common cause of damage to a fiber-optic cable during installation?

11. What are multifiber push-on (MPO) connectors?

12. Briefly describe Tier 1, Tier 2, Tier 3, and Tier 4 data centers, including their average annual downtimes.

13. What is the purpose of the free-floating ferrule in an MTP connector?

14. Explain why splicing is performed on optical fibers.

15. What is a fusion splice?

16. What is the main disadvantage of fusion splicing?

17. How are mechanical splices made?

18. What does polarity in a fiber-optic network refer to?

Chapter Review

19. Why should excess lengths of cable be avoided in an installation, even if those lengths are coiled and not creating a storage issue?

20. What is the most common width of an equipment rack?

21. What is a patch panel?

22. What is a distribution panel?

23. Why should VDV technicians always wear protective eyewear and other appropriate PPE when terminating fiber-optic cables and connectors?

24. Why is proper cable management more critical with fiber-optic cable than with copper cable?

25. Why are mounting holes typically spaced ⅝″ to 2″ apart in cable racks?

Chapter Activity Buffer Tube Color Codes

Many fiber-optic cables have multiple buffer tubes. The buffer tubes are color-coded per TIA-598-C. When the number of buffer tubes exceeds 12, the colors are repeated with the addition of a black tracer.

Write the color code abbreviation for each buffer tube as specified in the table.

Position Number	Base Color and Tracer	Abbreviation	Position Number	Base Color and Tracer	Abbreviation
1	Blue	BL	13	Blue with black tracer	D/BL
2	Orange	OR	14	Orange with black tracer	D/OR
3	Green	GR	15	Green with black tracer	D/GR
4	Brown	BR	16	Brown with black tracer	D/BR
5	Slate	SL	17	State with black tracer	D/SL
6	White	WH	18	White with black tracer	D/WH
7	Red	RD	19	Red with black tracer	D/RD
8	Black	BK	20	Black with yellow tracer	D/BK
9	Yellow	YL	21	Yellow with black tracer	D/YL
10	Violet	VI	22	Violet with black tracer	D/VI
11	Rose	RS	23	Rose with black tracer	D/RS
12	Aqua	AQ	24	Aqua with black tracer	D/AQ

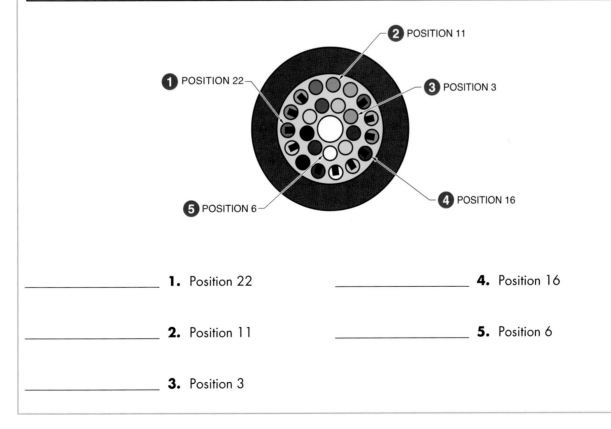

_____ **1.** Position 22   _____ **4.** Position 16

_____ **2.** Position 11   _____ **5.** Position 6

_____ **3.** Position 3

Chapter 12

Testing and Certification of Fiber-Optic VDV Cabling Systems

Unlike copper-based systems that use electricity, fiber-optic systems rely on light signals to transmit, receive, and process data. Because the quality of the system is dependent on the quality of the light transmission, VDV technicians must use specialized test equipment and procedures to test and certify fiber-optic installations. Because the presence of dust and dirt on a connector face can lead to poor test measurements, fiber-optic connectors must be properly cleaned. Damage to equipment and harm to personnel can be prevented by following proper safety practices. Fiber-optic strands are certified in order to receive the cable and connector manufacturers' product warranties.

- Describe how light transmission in a fiber-optic system is affected by attenuation.
- Explain how to measure attenuation.
- Describe how fiber-optic performance is affected by reflectance and encircled flux.
- Explain how light wavelengths relate to fiber-optic signal transmission.
- Describe different types of fiber-optic test equipment and which tests they are used for.
- Explain how to inspect and clean a fiber-optic connector.
- List the proper safety practices when testing and certifying fiber-optic systems.
- Explain how fiber-optic manufacturer product warranties are maintained in the industry.

Digital Resources

ATPeResources.com/QuickLinks
Access Code 838502

LIGHT TRANSMISSION

Light transmission is affected by attenuation. The less attenuation, the better the light transmission. As with testing and certification of copper-based systems, testing and certification of fiber-optic systems is performed for three reasons: verification of operation (which includes troubleshooting of nonoperational fibers), compliance with specifications, and certification of performance.

Verification of operation confirms that the correct fibers in a system are connected. It also confirms that a continuous path exists for the transmission of light signals. Compliance with specifications verifies that a system operates as designed and that each component, and the link as a whole, is within the loss budget. The performance of a fiber-optic system is measured as attenuation. Once a loss budget is developed, there is an expected amount of loss in a continuous fiber strand. Testing measures the attenuation for comparison with the loss budget. By performing tests against a loss budget, a VDV technician can verify that a cable was installed correctly and the system will function as intended. Finally, certification is performed to prove that a cable satisfies either an end user's requirements or, more commonly, the warranty requirements of the cable OEM.

SIGNAL MEASUREMENT

A *decibel (dB)* is a unit of measure used to express the strength of a signal or a gain or loss in optical or electrical power. For example, sound systems use the decibel as a unit of measure for volume (loudness). When a sound system is turned down to a low level, the measurement might be around -50 dB or -60 dB. As the volume increases, so does the value. At a level of 0 dB, the volume is extremely loud.

In the case of fiber-optic systems, a decibel is used to measure the difference in strength between a received signal and a standard signal source, or the amount of power lost when a pulse of light travels through a cable. Numbers on a decibel scale can be either positive or negative. When fiber-optic testing, the main concern is with

the loss of signal strength, so decibel multipliers are used to rate cable performance. For example, every 3 dB is equal to 50% of signal strength. A 6 dB loss equals a loss of 75% of signal strength. **See Figure 12-1. See Appendix.**

Fiber-Optic Decibel Multipliers		
Decibels	Loss Multiplier	Gain Multiplier
3	0.5	2
7	0.8	5
10	0.9	10

Figure 12-1. When fiber-optic testing, decibel multipliers are used to rate cable performance.

The operating range or acceptable loss depends on the strength of the signal source and the sensitivity of the receiver. Typically, an operating range of about 11 dB works. However, the acceptable loss for any fiber-optic link is determined by the loss budget. A loss budget is calculated by finding the sum of the individual acceptable loss events in a system. This loss budget must be less than the operating budget of the link. The loss and operating budgets are defined by the project's standards. **See Appendix.**

Although signal losses occur due to light dissipation or absorption along a cable's length, losses from splices and connectors often exceed the loss from the actual cable. Common signal losses occur from axial misalignment, angular misalignment, excessive end separation, and rough (unpolished) ends. **See Figure 12-2.**

REFLECTANCE

Reflectance, also known as return loss or back reflection, is an amount of light reflected back along the path of transmission from a connector or terminated fiber. Minimizing reflectance is important in modern fiber-optic systems. With standard, flat connectors, a small amount of light is reflected back towards the source, normally around -20 dB. Physical contact (PC) connectors have reflectance readings in the -30 dB to -40 dB range. Ultra-physical contact (UPC) connectors have reflectance readings in the −40 dB to −50 dB range. UPC connectors are similar to PC connectors but have a higher polish level, which further reduces reflectance.

another APC connector. They can withstand repeated plugging and unplugging without optical reflectance degradation. An APC connector directs any reflected light away from the core and out of the cable, so the lost light does not interfere with the primary signal. These connectors are not usually terminated in the field. They can be fusion spliced onto a cable or preordered as part of a plug-and-play system. APC connectors are typically used in high-speed cellular telephone systems. The reflected light can be seen as a small "bump" in a test instrument trace. **See Figure 12-3.**

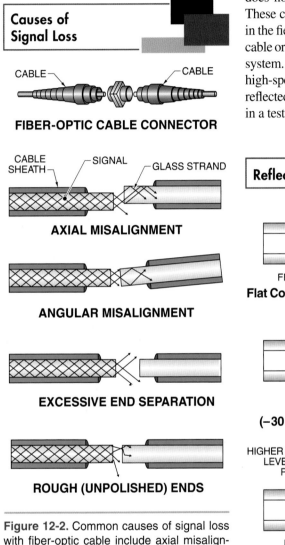

Figure 12-2. Common causes of signal loss with fiber-optic cable include axial misalignment, angular misalignment, excessive end separation, and rough (unpolished) ends.

Reflectance creates a small amount of interference with the primary signal being transmitted. Usually, this interference is so small that it is not a problem with signal integrity. However, in applications that require the full capability of the fiber, an angled physical contact connector can be used. An *angled physical contact (APC) connector* is a single-mode connector with an angled ferrule face that ensures low reflectance when joined with

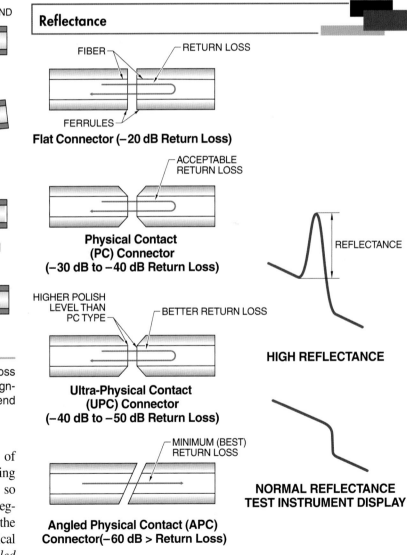

Figure 12-3. Reflectance is an amount of light reflected back along the path of transmission from a connector or terminated fiber and can be seen as a small "bump" in a test instrument trace.

ENCIRCLED FLUX

Encircled flux (EF) is the ratio between the transmitted power at a given distance from the center of the core of an optical fiber and the total power injected into the fiber. EF involves the amount of power coupled into the core of a fiber. If a light source is not aligned perfectly with the core of a fiber, an overfilled or underfilled condition can occur. Neither condition is desirable, but they have different consequences. **See Figure 12-4.**

An underfilled condition can result in a lower-than-actual attenuation measurement because the entire core loss is not being measured. An overfilled condition can result in a higher-than-actual loss measurement because some of the injected light is lost in the cladding. More precise testing cables and test equipment that condition the injected light more accurately are used to compensate for EF and provide more accurate test results. This is a recent development and is not yet readily accommodated in most existing test equipment. For a VDV technician, it is important to understand that EF is not an operational issue but rather one of compiling accurate test results. As testing for and accommodating EF conditions becomes more common, test equipment and procedures will become more automated. *Note:* EF-compliant testers are required as part of the ANSI/TIA-568-C standard.

Encircled Flux (EF)

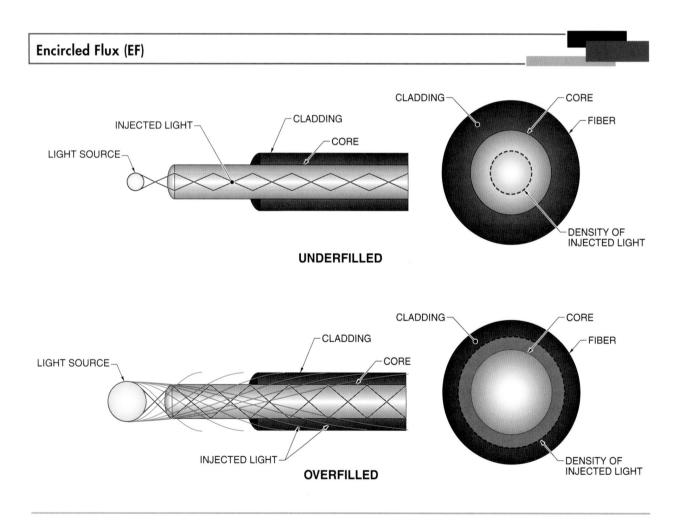

Figure 12-4. When a light source is not aligned perfectly with the core of a fiber, it creates an overfilled or underfilled condition called encircled flux.

WAVELENGTHS

Wavelength is the distance between two identical points in the adjacent cycles of an oscillating signal. Wavelength, rather than frequency, is used when describing fiber-optic system performance and is measured in nanometers (nm) or micrometers (mm). The light used in fiber-optic transmission is infrared and has a longer wavelength than the light visible to the human eye. Multimode and single-mode cables are constructed to have the lowest amount of attenuation at specific wavelengths. Accordingly, light sources are made to operate at those wavelengths. Fiber-optic systems must be tested and certified at those wavelengths as well. For this reason, tests are not run on a single strand of fiber. Instead, two tests with different wavelengths are usually run which correspond to the specific wavelengths optimized for the cable involved. For multimode fiber, 850 nm and 1300 nm are the wavelengths used, but they vary with single-mode fiber. The most common are 1310 nm and 1550 nm. **See Figure 12-5.** A VDV technician must know which wavelengths correspond to the cable being tested and set the tester accordingly.

FIBER-OPTIC SYSTEM QUALITY

Quality assurance (QA) is a planned and systematic set of actions necessary to provide adequate confidence that an item or product conforms to established technical requirements. *Quality control (QC)* is a system used to ensure that specified standards, including those for accuracy and quality, are met for manufactured products and materials. QA and QC for a fiber-optic installation is often performed by a qualified technician other than the installer. Whether by an installer or a QA/QC technician, installations must be inspected for quality per the following checklists. These tasks are performed for fiber-optic cabling on cable termination, system operation, and cable support, placement, and restraint.

> **Tech Tip**
>
> Dust, dirt, skin oil, and scratches can cause signal loss in a cable and result in "fail" readings on a fiber-inspection scope. Extreme cases of contamination result in no reading at all.

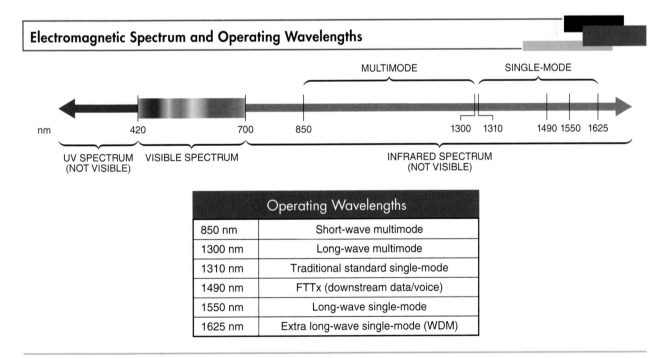

Electromagnetic Spectrum and Operating Wavelengths

Operating Wavelengths	
850 nm	Short-wave multimode
1300 nm	Long-wave multimode
1310 nm	Traditional standard single-mode
1490 nm	FTTx (downstream data/voice)
1550 nm	Long-wave single-mode
1625 nm	Extra long-wave single-mode (WDM)

Figure 12-5. Operating wavelengths in fiber optics are typically measured in nanometers (nm).

Some tasks involve QA and QC for cable termination. They include the following:

- Ensure that all cables, enclosures, and patch panels are properly labeled.
- Ensure that the end of the outer sheath is clean and sealed as required.
- Verify the order of fiber terminations by standard color code.
- Confirm that fibers are protected and placed properly in a tray or protected by use of a fan-out kit.
- Confirm that fiber connectors are inserted into termination points or protected with dust caps.
- Verify that all unpatched fiber ports on a patch panel or enclosure are covered with dust caps.
- Inspect all fiber-holding devices for both proper use and enclosure of each device per the manufacturer's recommendations. (For example, ensure splice trays are closed and secured, covers are in place, sliding shelves are secured in racks, and so on.)

Some tasks involve QA and QC for system operation. They include the following:

- Review the optical loss budget calculation and diagram.
- Verify the proper settings for test equipment and testing configuration.
- Review the test results to confirm that the operation of the system is within the loss budget.
- If OTDR tests have been performed, review the trace to confirm that the fiber strand conforms to the loss diagram and length expectations.
- Inspect and confirm that all boxes, hand holes, vaults, and so on, are properly closed.

Some tasks involve QA and QC for cable support, placement, and restraint. They include the following:

- Confirm that cable type, fiber type, and connector types meet specifications.
- Confirm that the cable type is correct for the installed environment, such as plenum, riser, or outdoor.
- Inspect the visible fiber pathway for minimum bend radius and for proper support without sharp transitions.

- Verify that the cable is supported no less than every 5′ or is in continuous cable tray.
- Verify that the crimp-tying devices are not compressing the outer sheath.

FIBER-OPTIC TESTING SAFETY

VDV technicians must always follow safety practices when testing fiber-optic cables. Safety practices include wearing appropriate eye protection and other PPE as required. VDV technicians must follow all federal, state, local, and job site safety regulations. They must verify that the light source is turned off at both ends of a cable and that the cable is disconnected at both ends. A VDV technician must never look directly into a fiber-optic connector or directly into lasers on communication or test equipment. Finally, they must verify that all unmated, fiber-optic connectors have a protective dust cap installed over the ferrule for each connector.

INSPECTING AND CLEANING FIBER-OPTIC CONNECTORS

VDV technicians must inspect and clean a fiber-optic connector when necessary before inserting it into a port or joining it with another connector. *Note:* This applies to all fiber-optic connectors including test jumpers, patch cords, launch cords, and receive cords. It is equally important to inspect the mating port or mating connector involved. A small amount of dirt, dust, or other contaminants on the face of a connector or in the interior of a port will block the cable's light signal. Contaminants on the face of a connector can be transferred to a port or to another connector. Sources of contamination include dust, lint, and skin oils. Contamination is the leading cause of fiber-optic cable failures.

It is impossible to know if the face of a connector is clean unless it is inspected. The face of a fiber-optic connector or port can be inspected with a fiber-inspection scope. The two main types of fiber-inspection scopes are optical fiber-inspection scopes and video fiber-inspection scopes. **See Figure 12-6.**

Fiber-Inspection Scopes

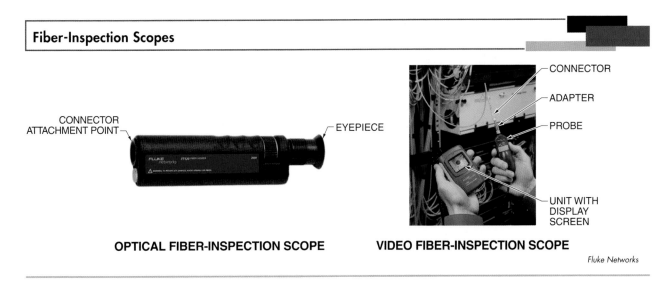

OPTICAL FIBER-INSPECTION SCOPE   VIDEO FIBER-INSPECTION SCOPE

Fluke Networks

Figure 12-6. The two main types of fiber-inspection scopes are optical fiber-inspection scopes and video fiber-inspection scopes.

Optical fiber-inspection scopes are handheld devices that use a lens to magnify the connector face image. The connector is attached to one end of the fiber inspection scope, and the VDV technician examines the connector face through an eyepiece. Adapters allow different types of connectors to be inspected. An optical fiber-inspection scope cannot be used to inspect ports.

WARNING: An optical fiber-inspection scope must only be used with a cable that is not connected to a light source, such as a loose cable. Viewing a cable that is connected to a light source with an optical fiber-inspection scope even once can cause permanent damage or blindness to the eye. Never use this type of instrument if the cable connections cannot be verified.

Video fiber-inspection scopes are handheld devices that consist of a probe and a separate unit with a display screen. The fiber-optic connector or port attaches to the probe, and a video image of the connector or port face is displayed on the screen. Adapters allow different types of connectors and connector ports to be inspected. Advanced video fiber-inspection scopes can evaluate a connector face, compare it to known standards,

and provide a "pass" or "fail" rating. This decreases the amount of time necessary to inspect the connector. This instrument is the most common type of instrument used to inspect fiber as it is safer than an optical fiber-inspection scope.

FIBER-OPTIC TEST EQUIPMENT TYPES

VDV technicians use different types of test equipment for fiber-optic cable installations. The tools range from simple, single-function units such as a visual fault locator to complex, multifunction units such as an optical time domain reflectometer (OTDR) with an integral visual fault locator or an optical loss test set (OLTS), which is a combination of a light source and a power meter. Some test tools are designed to only work with either single-mode or multimode cable. Other test tools are designed to work with either. The type of test tool used is job specific. In all cases, the VDV technician must read and follow the instructions for the specific tool in use.

One of the primary goals of testing a fiber-optic cable installation is to certify that it complies with known standards in order to obtain a warranty from the cabling OEM.

ANSI/TIA-568-C defines two tiers of testing fiber-optic cable. Tier 1 testing verifies the length and polarity of the cable, and measures the attenuation of the cable to verify that it is less than the loss budget. Tier 1 testing is typically performed with an optical attenuation test set or an OLTS and is considered the minimum testing level.

Tier 2 testing provides a graphical representation of the fiber-optic cable from end to end. The graph shows connectors, splices, and any faults if present. Tier 2 testing is performed with an OTDR and is useful in establishing a baseline picture of the cable.

There are different types of test equipment available for testing fiber-optic systems, and each has a slightly different menu and procedure for use. The main types of equipment used for testing fiber-optic systems include visual fault locators, optical attenuation test sets, OLTSs, and OTDRs. All fiber-optic test equipment operates by connecting the test instrument to one end of a fiber-optic strand. The physical connection is a matter of using a test lead that has the same type of connector and strand as the fiber being tested.

Proper cleaning of the fiber-optic strands is crucial to the success of a test procedure. Originally, an isopropyl alcohol-soaked wipe was recommended for cleaning fiber ends prior to testing any strand. It was discovered that over time, however, even an alcohol wipe deposits small amounts of residue which, after many applications, interferes with or hinders an optical signal. If an inspection reveals a contaminated fiber-optic connector, then it must be cleaned. Once it is cleaned, it must be inspected again. It may be necessary to clean a connector more than once to remove all contaminants.

There are a variety of products, both wet-cleaning and dry-cloth, that can be used to clean a connector face or port. Wet-cleaning products include lint-free wipes or swabs used with a cleaning fluid specifically designed for fiber-optic connectors. Some wipes or swabs are premoistened.

Most fiber-optic cable and connector OEMs now recommend using an improved cleaning tool known as a dry-cloth cleaning tool. Dry-cloth cleaning products include cleaning cassettes and pump-action cleaners. Cleaning cassettes have a dry, lint-free cloth inside the cassette body that cleans the connector face. Pump-action cleaners can be inserted into a fiber-optic connector or port. This is done with a simple pushing motion. **See Figure 12-7.** Pump-action cleaners are available in a variety of configurations for different connector and port types. Both cassettes and pump-action cleaners are preloaded with cleaning material that advances with each use. They have a limited number of uses after which they must be replaced.

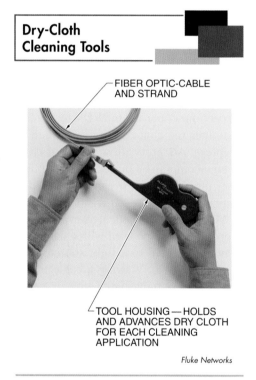

Dry-Cloth Cleaning Tools

FIBER OPTIC-CABLE AND STRAND

TOOL HOUSING — HOLDS AND ADVANCES DRY CLOTH FOR EACH CLEANING APPLICATION

Fluke Networks

Figure 12-7. With a dry-cloth cleaning tool, a small piece of clean cloth is rubbed over the glass end of a fiber strand to remove particulate matter.

Visual Fault Locators

A *visual fault locator* is a fiber-optic test tool that illuminates the locations of cable faults with the use of a bright red light source that escapes through faults in the cable jacket, splices, or connectors. A fiber-optic cable is connected to a visual fault locator using an integral adapter that accepts various types of fiber-optic connectors. The locator then emits a bright red light, either flashing or continuous, into the fiber-optic cable to identify faults such as broken fibers in cables and connectors, microbends, bad splices, and sharp bends. It can also be used to identify or locate fiber-optic cables, verify the continuity of fiber-optic cables, and verify the polarity of fiber-optic cables. Visual fault locators can also be used to verify that a reel of fiber-optic cable is undamaged before installing. **See Figure 12-8.**

or signal loss, in a fiber-optic cable. With an active light source at one end of a fiber-optic cable, the test set is connected to the other. To measure attenuation, the optical attenuation test set calculates the power in from the light source and subtracts the power from the power meter. The fiber-optic cable is tested at the wavelength in which it will operate. Some optical attenuation test sets have the ability to test two fiber-optic cables (a transmit fiber and a receive fiber) simultaneously. **See Figure 12-9.**

Visual Fault Locators

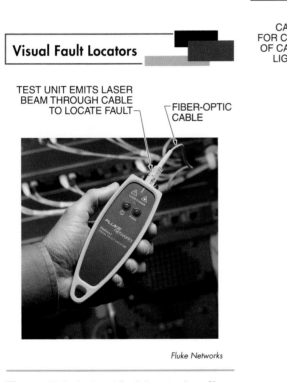

TEST UNIT EMITS LASER BEAM THROUGH CABLE TO LOCATE FAULT

FIBER-OPTIC CABLE

Fluke Networks

Figure 12-8. A visual fault locator is a fiber-optic test tool that illuminates the locations of cable faults with the use of a bright red light source that escapes through faults in the cable jacket, splices, or connectors.

Optical Attenuation Test Sets

An *optical attenuation test set* is a test instrument that is a power meter with cable inputs and is used to measure attenuation,

Optical Attenuation Test Sets

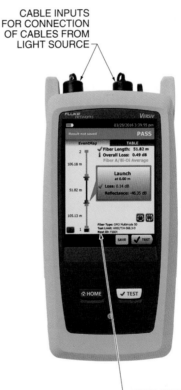

CABLE INPUTS FOR CONNECTION OF CABLES FROM LIGHT SOURCE

DISPLAY SCREEN INDICATES TEST PARAMETERS AND RESULTS

Fluke Networks

Figure 12-9. With an optical attenuation test set, the light source is connected to one end of a fiber-optic cable, and the power meter is connected to the other.

The test set is connected to the cable with a jumper, also known as a test-reference cord or reference jumper. Reference jumpers are provided with the test set and are high quality fiber-optic cables factory terminated at each end. The reference jumpers must match the cable and connector type being tested.

When testing multimode fiber-optic cable, if the test set is not encircled flux compliant, the light source (transmit) reference jumper must be wrapped around a mandrel. A *mandrel* is a fiber-wrapping device used to create attenuation within a fiber-optic cable. A mandrel is cylindrical in shape and is primarily used with multimode fiber-optic cable. The fiber is wrapped a certain number of turns around the mandrel, which counters the effects of the LED light source overfilling the fiber. **See Figure 12-10.**

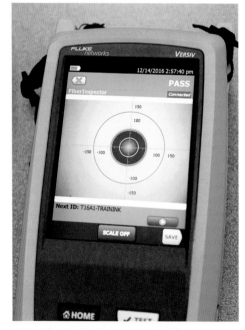

Video fiber-inspection scopes provide a safe view of a fiber end, allowing the technician to determine if the fiber can carry a signal without interference from dust, debris, or scratches.

Encircled Flux Compliant Test Equipment

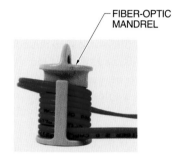

FIBER-OPTIC MANDREL

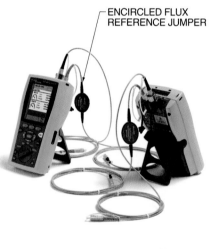

ENCIRCLED FLUX REFERENCE JUMPER

Fluke Networks

Figure 12-10. When testing multimode fiber-optic cable, if the test set is not encircled flux compliant, the light source (transmit) reference jumper must be wrapped around a mandrel.

The jumpers must be referenced before a Tier 1 test is conducted. Jumper referencing removes the jumper loss from the Tier 1 test. There are three methods of setting jumper references per the standards: the one-jumper reference method, the two-jumper reference method, and the three-jumper reference method. The reference method selected must correspond with the loss scenario of the link being tested. *Note:* The exact method of setting a reference and testing may vary between manufacturers of optical loss test sets. A VDV technician must always follow the test set manufacturer's instructions.

Using the one-jumper reference method, the light source and power meter are connected with one reference jumper. After the reference is established and recorded, the jumper is disconnected from the power meter. Using the jumper, the light source is then connected to one patch panel, and a second reference jumper is used to connect the power meter and another patch panel. The power reading is taken and recorded. The difference between the power reading and the reference is the loss. The loss includes the loss of the fiber-optic cable and both patch panels. **See Figure 12-11.**

Optical Attenuation Test with Single Reference Jumper

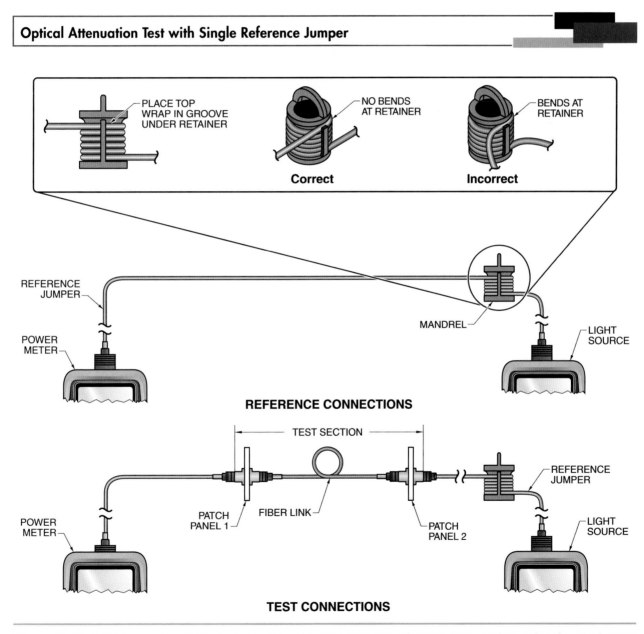

REFERENCE CONNECTIONS

TEST CONNECTIONS

Figure 12-11. A light source and power meter are connected with one reference jumper when using the one-jumper reference method.

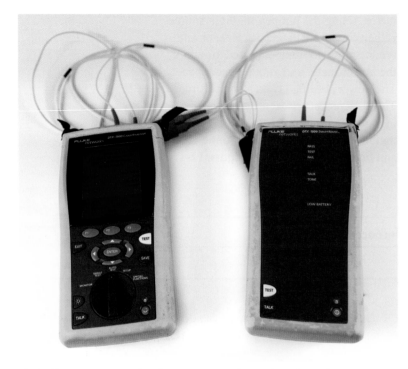

Optical loss test sets consist of separate transmitter and receiver units.

Using the two-jumper reference method, the light source and power meter are connected with two reference jumpers. After the reference is established and recorded, the two test jumpers are disconnected from each other, but they are not removed from the power meter or the light source. The light source test jumper is then connected to a patch panel, and the power meter test jumper is connected to the fiber-optic cable connector. The power reading is taken and recorded. The difference between the power reading and the reference is the loss. The loss includes the loss from the fiber-optic cable and the patch panel. **See Figure 12-12.**

Using the three-jumper reference method, the light source and power meter are connected with three reference jumpers. After the reference is established and recorded, the middle jumper is removed and not used as part of the test. The light source reference jumper is connected to one end of the fiber-optic cable, and the power meter reference jumper is connected to the other end. The power reading is

taken and recorded. The difference between the power reading and the reference is the loss. The loss only includes the loss of the fiber-optic cable. **See Figure 12-13.** Passing this test is typically the main requirement for certification and warranty of a fiber-optic system.

Previously, it was not difficult to purposely or inadvertently modify the parameters of tests performed with these devices. Improper use of these devices led to false results. As a general rule, it was not a good practice to have the same technicians who certified a fiber-optic system perform the final testing on that same system. Those who performed the initial certification could not always recognize the system's faults and would sometimes record inaccurate test results. This was mainly due to technicians having to manually record test results.

In order to prevent inaccuracies with final test results, a third-party testing and certification specialist was sometimes used to perform tests on a newly installed network to certify that it complied with published standards. By using a third party to certify an installation, expensive test equipment did not have to be purchased and money was saved. If a subcontractor was to install the system, a clause was typically added to the contract that stated that acceptance of the work is contingent upon the work being tested and approved by a qualified, independent, third-party testing service.

Currently, third-party testing is rarely used because modern test equipment electronically records test results. Errors are often nonexistent, and third-party testing is rarely, if ever, used. Third-party testing now adds an unnecessary expense to a project. Since the cost of test equipment has decreased over time, most technicians have access to the equipment required for most installations.

The proof of a system's functionality is in its operation, which is often a "yes" or "no" result. In the case of fiber-optic systems, they either work or they do not. The tests simply confirm power margin and expected loss.

Optical Attenuation Test with Two Reference Jumpers

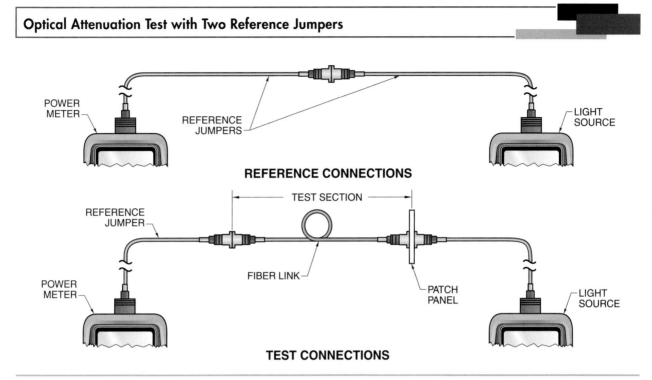

Figure 12-12. A light source and power meter are connected with two reference jumpers when using the two-jumper reference method.

Optical Attenuation Test with Three Reference Jumpers

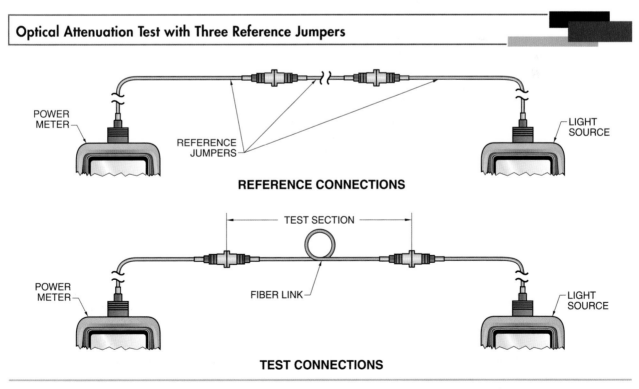

Figure 12-13. With the three-jumper reference method, the middle jumper is not used as part of the test and is only used to establish the reference.

Optical Loss Test Sets

An *optical loss test set (OLTS)* is a test instrument that is a combination of a light source and power meter used to detect and measure attenuation in a fiber-optic cable. OLTSs can be automated or manually operated. While automated units are typically more expensive than manual units, they are more accurate, less susceptible to human error, and have an easier training process. With an OLTS, the light source and power meter are calibrated together. By being calibrated together in the same manner, an OLTS is able to provide the most accurate readings possible. **See Figure 12-14.**

Optical Loss Test Sets (OLTSs)

OLTS CONTAINS LIGHT SOURCE AND POWER METER IN SAME UNIT

Fluke Networks

Figure 12-14. An OLTS is a test instrument that is a combination of a light source and a power meter used to detect and measure attenuation in a fiber-optic cable.

Optical Time Domain Reflectometers

An *optical time domain reflectometer (OTDR)* is a test instrument used to measure fiber-optic cable attenuation. It injects a pulse of laser light into a fiber-optic cable at a wavelength in which the fiber is intended to operate. The laser light is reflected back from locations along the fiber where there is a change in the index of refraction. These locations include connectors, splices, and damaged fiber. The reflected light is used to calculate the loss at each location and the distance to each location.

An OTDR can display test information as a graphical trace. The horizontal axis shows the distance, and the vertical axis shows the number of decibels. Locations where there is a change in the index of refraction appear as spikes along the trace. The OTDR can also display the information as a fiber map. Advanced OTDRs can evaluate a fiber being tested, compare it to a known standard, and provide a "pass" or "fail" rating. The results of each test can be stored in the OTDR and downloaded to a computer. Some OTDRs allow you to upload test results directly to the Internet. **See Figure 12-15.**

Similar to an optical loss test set, cables designed only for an OTDR are used to connect it to a fiber being tested. A *launch cord* is a fiber-optic cable used to introduce light from an optical source into another optical fiber. *A receive cord (tail cord)* is a fiber-optic cable used to measure insertion loss with an OTDR at the far end of a cable plant. The OTDR is connected to the fiber under test with a launch cord. The other (far) end is connected to a receive cord. These cables may also be referred to as launch and receive fibers.

The launch and receive cords are typically 100 m or longer in length and wound around a small ring or reel. Their minimum length can also be verified by consulting with the test instrument manufacturer. The launch and receive cords must match the cable type and connector type of the cable being tested. They are designed to compensate for the differences indicated and permit measurement of the first and last fiber-optic connector. **See Figure 12-16.**

Optical Time Domain Reflectometer (OTDR) Displays

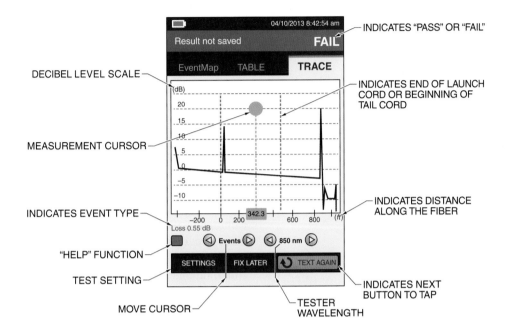

INDICATES "PASS" OR "FAIL"

DECIBEL LEVEL SCALE

INDICATES END OF LAUNCH CORD OR BEGINNING OF TAIL CORD

MEASUREMENT CURSOR

INDICATES EVENT TYPE

INDICATES DISTANCE ALONG THE FIBER

"HELP" FUNCTION

TEST SETTING

INDICATES NEXT BUTTON TO TAP

MOVE CURSOR

TESTER WAVELENGTH

OTDR GRAPHICAL TRACE

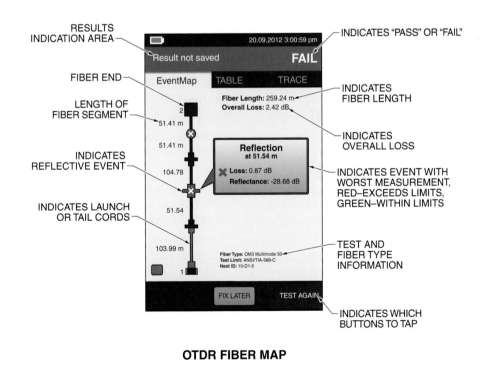

RESULTS INDICATION AREA

INDICATES "PASS" OR "FAIL"

FIBER END

INDICATES FIBER LENGTH

LENGTH OF FIBER SEGMENT

INDICATES OVERALL LOSS

INDICATES REFLECTIVE EVENT

INDICATES EVENT WITH WORST MEASUREMENT, RED–EXCEEDS LIMITS, GREEN–WITHIN LIMITS

INDICATES LAUNCH OR TAIL CORDS

TEST AND FIBER TYPE INFORMATION

INDICATES WHICH BUTTONS TO TAP

OTDR FIBER MAP

Figure 12-15. An OTDR can display test information as a graphical trace or as a fiber map.

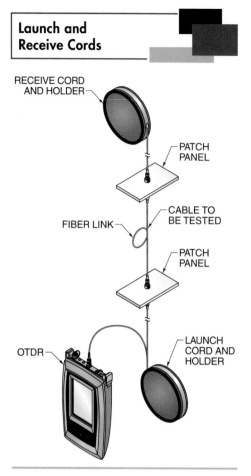

Launch and Receive Cords

RECEIVE CORD AND HOLDER

PATCH PANEL

FIBER LINK

CABLE TO BE TESTED

PATCH PANEL

OTDR

LAUNCH CORD AND HOLDER

Figure 12-16. Cables known as launch and receive cords are designed only for an OTDR and are used to connect it to a fiber being tested.

FIBER-OPTIC EQUIPMENT WARRANTIES

Most fiber-optic equipment manufacturers offer a 20-year or longer warranty. In order to maintain a warranty, manufacturers require test results for each fiber strand. The attenuation test is the most accurate and therefore the one usually requested. OTDR tests are sometimes requested by the end user because they provide more information about a fiber path and its composition, but they are not as accurate for operational measurements.

FIBER-OPTIC TECHNICAN CERTIFICATIONS

Building Industry Consulting Services International (BICSI) and the Fiber Optic

Association (FOA) have each established their own training and certification programs for VDV technicians. The programs are designed to ensure that VDV technicians have the knowledge and skills to successfully install VDV systems. Some job specifications may require the VDV technicians involved in a project to have a BICSI certification, an FOA certification, or both.

BICSI has four categories of cabling certification: Installer 1, Installer 2 Copper, Installer 2 Optical Fiber, and Technician. BICSI requires classroom instruction and passing a comprehensive exam in order to receive a certification. The topics a VDV technician must be proficient in to achieve these credentials include safety, professionalism, codes and standards, transmission fundamentals, media and connectors, structured cabling systems, bonding and grounding (earthing), installing support structures, pulling cable, firestopping systems, cable splicing, cable termination, testing and troubleshooting, TR/ER design, data centers, health care facilities, retrofits and upgrades, and planning and project management.

VDV technicians who wish to join the FOA and attain their credentials must demonstrate their knowledge, skills, and abilities (KSAs) in training courses or show experience in applying their KSAs in their work. The FOA Certified Fiber Optic Technician® (CFOT®) is the basic certification for most fiber-optic technicians. It is based on the KSAs deemed necessary for all technicians involved in the installation of fiber-optic networks and is recommended for anyone involved in the design or management of fiber-optic communications systems. The FOA also offers certification in premises cabling (CPCT,) which also includes copper wiring and wireless systems. The FOA also offers specialist certifications in three categories:

• Skills-based certifications are for CFOTs and are available in splicing, connectors, testing, OSP installation, and fiber characterization.

- Applications-based certifications (such as fiber-to-the-home (FTTH), local area networks (LANs), fiber-to-the-antenna (FTTA) and Data Centers) focus on the design and installation issues of each particular application and are aimed at a wider audience—users and managers overseeing these networks, designers and project managers, supervisors, and CFOTs involved in the installation. CFOT certifications are not a prerequisite for applications-based certifications, so a review of basic fiber optics for the course is available online as an introduction to the course.

- Fiber-optic network design is a specialist application certification intended for network owners, IT personnel, facilities managers, network designers, estimators, or technicians involved in the design or installation of fiber networks. This course is especially recommended for network owners and planners who may not be familiar with the process of fiber-optic network design, since it can make their jobs easier and help them improve their projects.

Tech Tip

Fiber-optic certification and training can be completed through programs offered by for-profit institutions, trade organizations, and community colleges as well as professional organizations such as BICSI and the FOA.

FOA members become certified by one of two methods. The first method is training. FOA certifications are available through FOA-approved schools that offer training that meets FOA standards. The programs are developed and delivered by experts in the fiber-optic industry, most of whom have over 20 years of experience as technicians, installers, manufacturers, and instructors. The second method is acquiring a Work-to-Cert direct certification based on industry experience. The FOA recognizes industry experience and has many members who qualified for direct certification based on their experience in the field. To acquire a Work-to-Cert direct certification, students must pass a comprehensive exam.

Summary

Fiber-optic performance is measured by the quality of the signal transmission and is tested for attenuation within the cable and at termination points. Common types of test equipment used to test and certify fiber-optic systems include visual fault locators, optical attenuation test sets, optical loss test sets and optical time domain reflectometers (OTDRs). In order to achieve accurate readings, fiber-optic connectors are properly cleaned by VDV technicians prior to taking test measurements. Because the glass in fiber-optic cables and connectors is difficult to see, VDV technicians must follow proper safety practices to prevent glass from getting on clothes, on hands, and possibly in the eyes. Equipment warranties can only be applied for after proper testing and certification of each fiber strand in the installation.

Chapter Review

1. List the three main reasons fiber-optic systems are tested and certified.

2. What criteria is used to measure the performance of a fiber-optic system?

3. In the case of fiber-optic systems, what is a decibel?

4. What is reflectance?

5. Explain the difference between physical contact (PC) and ultra-physical contact (UPC) connectors.

6. What is encircled flux?

7. Explain how an "underfilled" condition is different from an "overfilled" condition.

8. What is wavelength as related to fiber-optic systems?

9. What are the two tiers of fiber-optic cable testing as defined by standard ANSI/TIA-568-C?

10. Why are alcohol-soaked wipes no longer used for cleaning the ends of fiber-optic strands?

11. What is a visual fault locator?

12. What is the purpose of a mandrel when testing multimode fiber-optic cable?

Chapter Review

13. Why are launch and receive cords used with OTDRs?

14. What is length of the warranty offered by most fiber-optic equipment manufacturers?

15. What is the leading cause of fiber-optic cable failure?

Chapter Activity Fiber-Optic Test Equipment Purposes

Mark each empty box if the tool shown in the top row would be used for the purpose listed in the left-hand column.

Purpose	Optical Fiber-Inspection Scope	Video Fiber-Inspection Scope	Visual Fault Locator Light Source
Inspect a fiber-optic connector end for dirt and debris			
Provide a video image of a fiber-optic connector face			
Dry-clean a fiber-optic connector end			
Wet-clean of fiber-optic connector end			
Locate faults			
Check insertion loss			
Verify connectivity			
Verify polarity			
Acquire an in-depth evaluation of a fiber-optic cable			
Perform Tier 1 testing			
Perform Tier 2 testing			

OLTS	OTDR	Pump-Action Cleaner	Premoistened Cleaning Towels

Appendix

VDV System Abbreviations

EQUIPMENT

110	Twisted pair termination block
ADF	Area distribution facility
BDF	Building distribution frame or facility
BEF	Building entrance frame or facility
CAB	Telecom cabinet or enclosure
CONN	Connector
CSC	Copper splice closure
CVE	Controlled environmental vault
ER	Equipment room
FDF	Fiber distribution facility
FF	Front facing
FS	Fiber shelf/fiber termination panel
FSC	Fiber-optic splice enclosure
HH	Handhole
IDC	Internet data center
IDF	Intermediate distribution frame or facility
MDF	Main distribution frame
MH	Manhole; maintenance hole
MPOE	Minimum point of entry; main point of entry
OCEF	Optical cable entrance facility
OTDR	Optical time domain reflectometer
PAV	Pavement
PBB	Primary bonding busbar
PC	Plastic conduit
PG	Pair group
POP	Point of presence
PR	Pair
PVC	Polyvinyl chloride
R/W	Right-of-way
SBB	Secondary bonding busbar
SC	Splice closure
SCS	Structured cabling system
SER	Serial
SMR	Surface-mounted raceway
SS	Fiber splice shelf
TEBC	Telecom equipment bounding busbar
TC	Telecom closet
TCC	Telecom conduit
TCH	Telecom conduit sleeve, horizontal
TCR	Telecom horizontal and vertical riser conduit
TCT	Telecom cable tray

EQUIPMENT

TEC	Telecom entrance conduit
TEL	Telephone
TELECOM	Telecommunications
TERM	Terminal
TP	Twisted pair
TPB	Telecom pull box
TR	Telecom room
TSL	Telecom wall or floor slot
TSV	Telecom conduit sleeve, vertical

WIRE AND CABLE

AFMW	24 AWG bonded fill flooded twisted cable
ARMM	24 AWG riser armored bonded multipair cable
AWG	American wire gauge
CAT3	Category 3 twisted pair copper cable
CAT6	Category 6 twisted pair copper cable
CM	Communications cable
CMP	Communications plenum cable
CMR	Communications riser cable
COAX	Coaxial cable
FO	Fiber optic
HDPE	High-density polyethylene
LTFF	Loose tube filled and flooded
MDPE	Medium-density polyethylene
MM	Loose tube filled and flooded
MPP	Multipurpose plenum cable
OFC	Optical fiber conductive cable
OFCP	Optical fiber conductive plenum cable
OFCR	Optical fiber conductive riser cable
OFNR	Optical fiber nonconductive riser cable
OFN	Optical fiber nonconductive cable
OFNP	Optical fiber nonconductive plenum cable
SM	Single-mode fiber-optic cable
STP	Shielded twisted pair
TB	Tight buffered
TBC	Telecom bounding cable
UTP	Unshielded twisted pair

Commonl Industry Abbreviations for Cable Construction

Industry Acronyms	ISO/IEC 11801 Name	Cable Shielding	Pair Shielding
UTP	U/UTP	none	none
STP, ScTP	–	none	foil
FTP, STP, ScTp	–	foil	none
STP, ScTP	S/UTP	braiding	none
SFTP, S-FTP, STP	SF/UTP	braiding, foil	none
FFTP	F/FTP	foil	foil
SSTP, SFTP, STP	S/FTP	braiding	foil
SSTP, SFTP	SF/FTP	braiding,foil	foil

The code before the slash designates the shielding for the cable itself, while the code after the slash determines the shielding for the individual pairs:

U = Unshielded
F = Foil shielded
S = Braided shielding (outer layer only)
TP = Twisted Pair

TIA 568A and 568B Wiring Color Codes

TIA 568A		TIA568A	
Pin #	Wire Color	Pin#	Wire Color
1	White/Green	1	White/Orange
2	Green	2	Orange
3	White/Orange	3	White/Green
4	Blue	4	Blue
5	White/Blue	5	White/Blue
6	Orange	6	Green
7	White/Brown	7	White/Brown
8	Brown	8	Brown

ANSI/TIA Telecommunications Standards

- Generic Telecommunications Cabling for Customer Premises (ANSI/TIA-568.0-D)
- Commercial Building Telecommunications Cabling Standard (ANSI/TIA-568.1-D)
- Balanced Twisted-Pair Telecommunications Cabling and Components Standard (ANSI/TIA-568-C.2)
- Optical Fiber Components Standard (ANSI/TIA-568-C.3)
- Broadband Coaxial Cabling and Components Standard (ANSI/TIA-568-C.4)
- Telecommunications Pathways and Spaces (ANSI/TIA-569-D)
- Residential Telecommunications Infrastructure Standard (ANSI/TIA-570-C)
- Administration Standard for Telecommunications Infrastructure (ANSI/TIA-606-B)
- Generic Telecommunications Bonding and Grounding (Earthing) for Customer Premises (ANSI/TIA-607-C)
- Customer-Owned Outside Plant Telecommunications Infrastructure Standard (ANSI/TIA-758-B)
- Structured Cabling Infrastructure Standard for Intelligent Building Systems (ANSI/TIA-862-B)
- Telecommunications Infrastructure Standard for Data Centers (ANSI/TIA-942-A)
- Telecommunications Infrastructure Standard for Industrial Premises (ANSI/TIA-1005-A)
- Healthcare Facility Telecommunications Infrastructure Standard (ANSI/TIA-1179)
- Telecommunications Infrastructure Standard for Educational Facilities (ANSI/TIA-4966)

Common Prefixes

Symbol	Prefix	Equivalent
G	giga	1,000,000,000
M	mega	1,000,000
k	kilo	1000
base unit	—	1
m	milli	0.001
μ	micro	0.000001
n	nano	0.000000001
p	pico	0.000000000001

Metric Prefixes

Multiples and Submultiples	Prefixes	Symbols	Meanings
$1,000,000,000,000 = 10^{12}$	tera	T	trillion
$1,000,000,000 = 10^{9}$	giga	G	billion
$1,000,000 = 10^{6}$	mega	M	million
$1000 = 10^{3}$	kilo	k	thousand
$100 = 10^{2}$	hecto	h	hundred
$10 = 10^{1}$	deka	d	ten
Unit $1 = 10^{0}$			
$0.1 = 10^{-1}$	deci	d	tenth
$0.01 = 10^{-2}$	centi	c	hundredth
$0.001 = 10^{-3}$	milli	m	thousandth
$0.000001 = 10^{-6}$	micro	μ	millionth
$0.000000001 = 10^{-9}$	nano	n	billionth
$0.000000000001 = 10^{-12}$	pico	p	trillionth

Metric Conversions

Initial Units	Final Units											
	giga	mega	kilo	hecto	deka	base unit	deci	centi	milli	micro	nano	pico
giga		3R	6R	7R	8R	9R	10R	11R	12R	15R	18R	21R
mega	3L		3R	4R	5R	6R	7R	8R	9R	12R	15R	18R
kilo	6L	3L		1R	2R	3R	4R	5R	6R	9R	12R	15R
hecto	7L	4L	1L		1R	2R	3R	4R	5R	8R	11R	14R
deka	8L	5L	2L	1L		1R	2R	3R	4R	7R	10R	13R
base unit	9L	6L	3L	2L	1L		1R	2R	3R	6R	9R	12R
deci	10L	7L	4L	3L	2L	1L		1R	2R	5R	8R	11R
centi	11L	8L	5L	4L	3L	2L	1L		1R	4R	7R	10R
milli	12L	9L	6L	5L	4L	3L	2L	1L		3R	6R	9R
micro	15L	12L	9L	8L	7L	6L	5L	4L	3L		3R	6R
nano	18L	15L	12L	11L	10L	9L	8L	7L	6L	3L		3R
pico	21L	18L	15L	14L	13L	12L	11L	10L	9L	6L	3L	

Watts to Decibel Conversion

Watts	dBm ‡
100*	20
20*	13
10*	10
8*	9
5*	7
4*	6
2*	3
1*	0
500†	–3
250†	–6
200†	–7
125†	–9
100†	–10
50†	–13
25†	–16
20†	–17
12.5†	–19
10†	–20
5†	–23
1†	–30
500§	–33

* in mW
† in µW
‡ in dBm
§ in nW

Network Device Symbols

Network devices are represented in network diagrams by symbols that indicate their respective functions. The telecommunications industry has not standardized these symbols, but many are used so commonly that they have become de facto standards. Variations are commonly used to denote devices with special features or a combination of network functions.

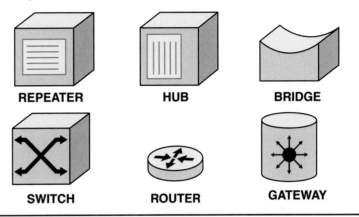

REPEATER   HUB   BRIDGE

SWITCH   ROUTER   GATEWAY

Maximum Work Area Cable Length

Length of Horizontal Cable*	Maximum Length of Work Area Cable*	Maximum Combined Length of all Patch and Equipment Cords*
295	16	33
279	30	46
262	44	59
246	57	72
230	72	89

* in ft

Copper Minimum Bend Radius

Cable Type	Bend Radius
4 Pair UTP	4 × cable diameter
4 Pair ScTp	8 × cable diameter
Backbone	10 × cable diameter
Patch Cords	Under Review

Maximum Backbone Distance

Cable Type	Main to Horizontal Cross Connect	Main to Intermediate Cross Connect	Intermediate to Horizontal Cross Connect
Copper (Voice)	2624	1640	984
Multimode Fiber	6560	5575	984
Singlemode Fiber	9840	8855	984

* in ft

Fiber-Optic Minimum Bend Radius

	No Load Condition	Maximum Load
Intrabuilding 2 or 4 Fiber	25 mm	50 mm
Intrabuilding Backbone	10 × OD	15 × OD
Interbuilding Backbone	10 × OD	20 × OD

Optical Test Equipment Summary

Type of Test	Equipment	Investment	Accuracy	Customer Requirements	To Be Used For
Visual	Visual Fault Locator	$300-$400	Good	Never	Continuity, Polarity
Attenuation	Optical Test Set	$4K- $8K	Better	Always	Actual Link Loss, Continuity, Polarity
Faults	OTDR	$8K-$17K	Best	On Occasion	Signature Traces, Length Measurements Events Analysis (Splice & Connector) Troubleshooting

Optical Fiber Cable Transmission Performance

Cable Type*	Cable Type Wavelength†	Maximum Attenuation‡	Minimum Transmission Capacity§
50/125 Multimode	850	3.5	500
	1300	1.5	500
62.5/125 Multimode	850	3.5	160
	1300	1.5	500
Singlemode inside	1310	1.0	N/A
Plant Cable	1550	1.0	N/A
Singlemode Outside Plant Cable	1310	0.5	N/A
	1550	0.5	N/A

Note: The manufacturer's documentaion on the fiber's performance can be used to demondtrate compliance with the above performance requirements.

* in µm
† in nm
‡ in db/km
§ in Mhz × km

Determining Loss Budgets*

Component	Qty./Unit	Unit Loss from Specification Multiplier	Loss†
Cable/Loss	3 / km	3.0 dB/km	9
Connector Loss	2 / connectors	0.75 dB/connector	1.5
Splice Loss	1/ splice	0.3 dB/splice	0.3
Total Loss			10.8

*3 km, 50 µm system with 2 connector pairs and a splice at 850 nm
Note: Cable loss will vary with operating wavelengths of 850nm, 1300 nm, 1310 nm, and 1550 nm
† in dB

Miscellaneous VDV Symbols

Device	Abbrev	Symbol	Device	Abbrev	Symbol
Cable Antenna Outlet (CATV)	CAO	TV	Plywood Backboard		
Master Antenna Outlet (MATV)	MAO	TV M	Telephone Undercarpet Flat Conductor	UCT	---UCT---
Equipment Cabinet	EQC		Data Undercarpet Flat Conductor	UCD	---UCD---
Equipment Rack	EQR	Wall Mounted	Power Pole	PP	P ■ 2 POWER CIRCUIT
Equipment Rack	EQR	Freestanding	Telecom Pole	TP	T ■
Terminal Cabinet	TC	TCC	Telecom Power Pole	TPP	TP ■ 2 POWER CIRCUIT

Electrical/Electronic Abbreviations/Acronyms . . .

Abbr/Acronym	Meaning	Abbr/Acronym	Meaning
A	Ammeter; Ampere; Anode; Armature	DPST	Double-Pole, Single-Throw
AC	Alternating Current	DS	Drum Switch
AC/DC	Alternating Current; Direct Current	DT	Double-Throw
A/D	Analog to Digital	DVM	Digital Voltmeter
AF	Audio Frequency	EMF	Electromotive Force
AFC	Automatic Frequency Control	F	Fahrenheit; Fast; Forward; Fuse; Farad
Ag	Silver	FET	Field-Effect Transistor
ALM	Alarm	FF	Flip-Flop
AM	Ammeter; Amplitude Modulation	FLC	Full-Load Current
AM/FM	Amplitude Modulation; Frequency Modulation	FLS	Flow Switch
ARM.	Armature	FLT	Full-Load Torque
Au	Gold	FM	Frequency Modulation
AU	Automatic	FREQ	Frequency
AVC	Automatic Volume Control	FS	Float Switch
AWG	American Wire Gauge	FTS	Foot Switch
BAT.	Battery (electric)	FU	Fuse
BCD	Binary-Coded Decimal	FWD	Forward
BJT	Bipolar Junction Transistor	G	Gate; Giga; Green; Conductance
BK	Black	GEN	Generator
BL	Blue	GRD	Ground
BR	Brake Relay; Brown	GY	Gray
C	Celsius; Capacitance; Capacitor, Coulomb	H	Henry; High Side of Transformer; Magnetic Flux
CAP.	Capacitor	HF	High Frequency
CB	Circuit Breaker; Citizens Band	HP	Horsepower
CC	Common-Collector; Configuration	Hz	Hertz
CCW	Counterclockwise	I	Current
CE	Common-Emitter Configuration	IC	Integrated Circuit
CEMF	Counter-Electromotive Force	INT	Intermediate; Interrupt
CKT	Circuit	IOL	Instantaneous Overload
cmil	Circular Mil	IR	Infrared
CONT	Continuous; Control	ITB	Inverse Time Breaker
CPS	Cycles Per Second	ITCB	Instantaneous Trip Circuit Breaker
CPU	Central Processing Unit	J	Joule
CR	Control Relay	JB	Junction Box
CRM	Control Relay Master	JFET	Junction Field-Effect Transistor
CT	Current Transformer	K	Kilo; Cathode
CW	Clockwise	kWh	kilowatt-hour
D	Diameter; Diode; Down	L	Line; Load; Coil; Inductance
D/A	Digital to Analog	LB-FT	Pounds Per Foot
DB	Dynamic Braking Contactor; Relay	LB-IN.	Pounds Per Inch
DC	Direct Current	LC	Inductance-Capacitance
DIO	Diode	LCD	Liquid Crystal Display
DISC.	Disconnect Switch	LCR	Inductance-Capacitance-Resistance
DMM	Digital Multimeter	LED	Light-Emitting Diode
DP	Double-Pole	LRC	Locked Rotor Current
DPDT	Double-Pole, Double-Throw	LS	Limit Switch

. . . Electrical/Electronic Abbreviations/Acronyms

Abbr/ Acronym	Meaning	Abbr/ Acronym	Meaning
LT	Lamp	RF	Radio Frequency
M	Motor; Motor Starter; Motor Starter Contacts	RH	Rheostat
MAX.	Maximum	rms	Root-Mean-Square
MB	Magnetic Brake	ROM	Read-Only Memory
MCS	Motor Circuit Switch	rpm	Revolutions Per Minute
MEM	Memory	RPS	Revolutions Per Second
MED	Medium	S	Series; Slow; South; Switch; Second; Siemen
MIN	Minimum	SCR	Silicon-Controlled Rectifier
MMF	Magnetomotive Force	SEC	Secondary
MN	Manual	SF	Service Factor
MOS	Metal-Oxide Semiconductor	1 PH; 1ϕ	Single-Phase
MOSFET	Metal-Oxide Semiconductor Field-Effect Transistor	SOC	Socket
MTR	Motor	SOL	Solenoid
N; NEG	North; Negative; Number of Turns	SP	Single-Pole
NC	Normally Closed	SPDT	Single-Pole, Double-Throw
NEUT	Neutral	SPST	Single-Pole, Single-Throw
NO	Normally Open	SS	Selector Switch
NPN	Negative-Positive-Negative	SSW	Safety Switch
NTDF	Nontime-Delay Fuse	SW	Switch
O	Orange	T	Tera; Terminal; Torque; Transformer
OCPD	Overcurrent Protection Device	TB	Terminal Board
OHM	Ohmmeter	3 PH; 3ϕ	Three-Phase
OL	Overload Relay	TD	Time-Delay
OZ/IN.	Ounces Per Inch	TDF	Time-Delay Fuse
P	Peak; Positive; Power; Power Consumed	TEMP	Temperature
PB	Pushbutton	THS	Thermostat Switch
PCB	Printed Circuit Board	TR	Time-Delay Relay
PH	Phase	TTL	Transistor-Transistor Logic
PLS	Plugging Switch	U	Up
PNP	Positive-Negative-Positive	UCL	Unclamp
POS	Positive	UHF	Ultrahigh Frequency
POT.	Potentiometer	UJT	Unijunction Transistor
P-P	Peak-to-Peak	UV	Ultraviolet; Undervoltage
PRI	Primary Switch	V	Violet; Volt
PS	Pressure Switch	VA	Voltampere
PSI	Pounds Per Square Inch	VAC	Volts Alternating Current
PUT	Pull-Up Torque	VDC	Volts Direct Current
Q	Transistor; Quality Factor	VHF	Very High Frequency
R	Radius; Red; Resistance; Reverse	VLF	Very Low Frequency
RAM	Random-Access Memory	VOM	Volt-Ohm-Milliammeter
RC	Resistance-Capacitance	W	Watt; White
RCL	Resistance-Inductance-Capacitance	w/	With
REC	Rectifier	X	Low Side of Transformer
RES	Resistor	Y	Yellow
REV	Reverse	Z	Impedance

Industry and Standards Organizations . . .

Factory Mutual (FM) www.fmglobal.com	Tests equipment and products to verify conformance to national codes and standards
International Association of Electrical Inspectors (IAEI) www.iaei.org	Focuses on interpretation of the National Electrical Code® and teaches safe installation and use of electricity
International Code Council (ICC) www.iccsafe.org	Administers the International Building Code® including building materials, building systems, energy efficiency, fire protection systems, and structural design
National Electrical Installation Standards (NEIS) National Electrical Contractors Association (NECA) www.neca-neis.org	Standards developed by NECA in partnership with other industry organizations; all NEIS are submitted for approval to ANSI
National Electrical Manufacturers Association (NEMA) www.nema.org	Provides standards for manufacturers of electrical equipment
InterNational Electrical Testing Association (NETA) www.netaworld.org	Defines standards by which electrical equipment is deemed safe and reliable; also provides training for and certification of electrical testing technicians
International Electrotechnical Commission (IEC) www.iec.ch	Provides standards for most international installations and some domestic installations
National Fire Protection Association (NFPA) www.nfpa.org	Provides guidance in assessing hazards of products of combustion; publishes the National Electrical Code®; develops hazardous materials information
American National Standards Institute (ANSI) www.ansi.org	Coordinates and encourages activities in national standards department; identifies industrial and public needs for standards; acts as national coordinator and clearinghouse for consensus standards
The Institute of Electrical and Electronics Engineers (IEEE) www.ieee.org	Provides guidance in all electrical and electronic systems including aerospace systems, computers and telecommunications, biomedical engineering, electric power, and consumer electronics
Mine Safety and Health Administration (MSHA) www.msha.gov	Develops and enforces safety and health rules for mines and for certain equipment that may be used in other locations
Occupational Safety and Health Administration (OSHA) www.osha.gov	Develops and enforces safety and health rules for most public and private workplaces; provides training, outreach, education, and assistance

. . . Industry and Standards Organizations	
National Institute for Occupational Safety and Health (NIOSH) www.cdc.gov/niosh	Acts in conjunction with OSHA to develop recommended exposure limits for hazardous substances or conditions located in the workplace; recommends preventive measures to reduce or eliminate adverse health and safety effects
Underwriters Laboratories Inc. (UL) www.ul.com	Tests equipment and products to verify conformance to national codes and standards
Canadian Standards Association (CSA) www.csa.ca	Tests equipment and products to verify conformance to Canadian codes and standards

Glossary

A

acceptance angle: The maximum angle at which light can enter a glass fiber core as an input and continue to reflect off the boundary layer between the core and the cladding.

access control system: A security system that controls door locks.

A-frame ladder: *See* stepladder.

alternating current equipment ground (ACEG): A conductor that bonds an electrical panel to a bonding busbar.

American National Standards Institute (ANSI): A national organization that helps identify industrial and public needs for standards.

angled physical contact (APC) connector: A single-mode connector with an angled ferrule face that ensures low reflectance when joined with another APC connector.

aramid: A high-strength, flexible, heat-resistant synthetic fiber. Also known as Kevlar®.

armored fiber-optic cable: A fiber-optic cable that utilizes an interlocking metallic armored design which eliminates the need to install rigid conduit to meet building codes.

attenuation: The reduction of power in any signal, light beam, or light wave, either completely or as a percentage of a reference value.

attenuation to crosstalk ratio (ACR): The measure (ratio) of signal loss (attenuation) to near-end crosstalk.

attenuation to crosstalk ratio far-end (ACR-F): The ratio of signal strength to undesired signal noise in a twisted-pair cable, as measured at the far end of the cable. Also known as equal level far-end crosstalk (ELFEXT).

audio-visual (AV) system: A network of hardwired or wireless video monitors that transmits sound and video, usually for presentations to large groups of people located in a common area.

authority having jurisdiction (AHJ): A person who has the delegated authority to determine, mandate, and enforce code requirements established by jurisdictional governing bodies.

B

backbone bonding conductor (BBC): A conductor that connects multiple telecommunications bonding backbone conductors.

backbone cabling: Telecommunications cabling and hardware that runs from a telecommunications closet, or intermediate distribution frame (IDF), to a main distribution frame (MDF) or to another telecommunications room. Also known as riser cabling.

back reflection: *See* reflectance.

bandwidth: The amount of data that can be sent through a given channel and is measured in bits per second (bps).

bend radius: The minimum radius that a cable can bend without causing damage.

binder: Colored plastic tape or colored string-like fiber that is wrapped helically around a specific group of 25 twisted pairs.

blown fiber: An installation method in which a housing containing many smaller tubes is installed without fiber.

bonding: The act of connecting two metallic parts to form a continuous, low-impedance electrical path.

bonding conductor: A conductor used to bond metal objects when required. Also known as a bonding jumper.

bonding jumper: *See* bonding conductor.

bridging clip: A small piece of steel that snaps over two center IDCs and thus connects (bridges) the two outer IDCs.

buffer: A fiber-optic cable layer that provides protection for optical fiber within a cable.

Building Industry Consulting Service International (BICSI): A professional telecommunications standards and certification organization.

C

cable: A group of two or more conductors within a common protective cover and is used to connect individual components.

cable loss report: An analytical tool that can be created by certain types of test instruments and is usually generated by a VDV technician to certify all of the fibers in a specific cable.

cable sheath: An outer layer used to provide protection for a cable with multiple fiber strands.

campus cabling: Telecommunications premise cabling that interconnects buildings via underground or overhead cabling means.

certification tester: A test instrument used to test and certify cables in a system in order to obtain a warranty from the structured cabling system OEM.

channel: The end-to-end transmission or communications path over which application-specific equipment is connected.

chromatic dispersion: The slowing of bandwidth caused by the spread of a light pulse due to the varying refraction rates of the different colored wavelengths.

cladding: A layer of glass fused to a glass fiber core to aid in performance of light transmission and act as protection for the core.

coating: A layer of protective material applied to a fiber-optic glass strand.

coaxial cable: A low-voltage cable comprised of an insulated central conducting wire wrapped within an additional cylindrical shield.

conductive cable: A cable that contains a metal component of some type, which would be subject to induced electrical current in the event of a lightning strike or physical contact with an energized electrical component.

conductor: A low-resistance metal that carries electricity to various parts of a circuit.

connector: A device that connects wires or fibers in a cable to equipment or other wires and fibers.

Construction Specifications Institute (CSI): An organization that develops standardized construction specifications.

controller: A device used to control another piece of VDV equipment.

cord: A group of two or more flexible conductors within a common protective cover that is used to deliver power to a load by means of a plug.

core: The smallest section of a glass fiber strand and is used to carry light waves.

crimp-on connector: A fiber-optic connector that is crimped together with a crimping tool to complete the mechanical splice inside the connector.

cross-connect: A facility enabling the termination of network cables as well as their interconnection or cross-connection with other cables or devices.

crossed pair: A fault that occurs when the termination positions of two pairs are transposed, or connected to the wrong pair, at one end of a VDV cable.

crossed wire: A fault that occurs when the termination positions of two VDV conductors are transposed, or connected to the wrong conductor, at one end of a cable.

crosstalk: A type of interference caused by signals from one circuit into adjacent circuits.

D

data center: A building or part of a building that is designed, built, and maintained specifically to house and connect computer servers.

decibel (dB): 1. A unit of measure used to express the relative intensity of sound. **2.** A unit of measure used to express the strength of a signal or a gain or loss in optical or electrical power.

delay skew: The signal delay difference between the fastest and slowest pair in a twisted-pair cable. Also known as propagation delay.

E

earmuffs: A hearing protection device worn over the ears.

earplug: A hearing ear-protection device made of moldable synthetic rubber or plastic foam that is inserted into the ear canal.

electromagnetic interference (EMI): Interference in signal transmission or reception caused by the radiation from electric and magnetic fields.

electrostatic discharge (ESD): A sudden movement of electricity between two objects.

encircled flux (EF): The ratio between the transmitted power at a given distance from the center of the core of an optical fiber and the total power injected into the fiber.

equal level far-end crosstalk (ELFEXT): *See* attenuation to crosstalk ratio far-end (ACR-F).

equipment room (ER): A central space that serves telecommunications equipment for occupants of a building or a campus of several buildings.

F

face shield: An eye-and-face protection device that covers the wearer's entire face with a plastic shield.

far-end crosstalk (FEXT): The measure of signal interference at the transmit end of a twisted-pair cable from a transmit pair to an adjacent pair, as measured at the far end of the cable.

fiber-optic cable: A strand composed of pure glass that is used to carry digital signals, in the form of modulated light, over long distances.

fiber-optic fusion splicer: A tool used to weld the ends of two strands of fiber into a single length using an electrical arc to melt the ends of the strands.

fiber-optic termination kit: A tool kit used by a VDV technician that contains all of the necessary tools for terminating fiber-optic pre-polished and field-polished connectors.

fiber scope: A device with a magnifying lens on one end and a fiber-optic connection on the other.

field-polish connector: A fiber-optic connector that requires hand-polishing of the glass protruding from the end of the connector ferrule. Also known as a hand-polish connector.

fire alarm system: A low-voltage system used to notify building occupants of a possible fire upon the detection of smoke, heat, or both.

firestop: A system made of various fire-rated components used to inhibit the spread of fire.

firestopping: 1. The process of applying and installing a material or member that seals an opening in a fire-rated wall, floor, or membrane to inhibit the spread of fire, smoke, and fumes in a structure. **2.** The process of sealing any penetration in a fire-rated building element in order to maintain its rating.

firestopping caulk: A material commonly used to seal holes drilled through walls or structural members to prevent the spread of fire.

fish tape: A flexible cable that is used to move another cable through a tight space that cannot be reached.

floor plan: A drawing that gives a plan view of each floor of a building.

fusion splice: A permanent splice accomplished by the application of heat with a temperature high enough to melt the ends of two sections of glass fiber.

fusion splicer: A device that uses a high-voltage electrical arc between two electrodes to heat and melt the ends of glass fibers in order to join them and form a continuous strand.

G

global positioning system (GPS): A worldwide positioning system that uses signals from navigational satellites orbiting the Earth to determine the location and elevation of a GPS receiver.

glow rod: A semirigid rod that is used to move a cable through a space that cannot be reached.

goggles: An eye-protection device with a flexible frame that is secured to the wearer's face with an elastic headband.

grounding: A low-resistance conducting connection between electrical circuits, equipment, and the earth.

grounding electrode: A conductive metal object used to establish a connection from an electrical circuit to the earth.

grounding electrode conductor (GEC): A conductor that connects parts of an electrical distribution system (equipment grounding conductors, grounded conductors, and all metal parts) to an approved grounding electrode system.

grounding electrode system: A system of all the grounding conductors in a building or structure that are bonded together.

H

hand-polish connector: *See* field-polish connector.

high-visibility clothing: A class of clothing made with extremely bright, color-enhanced fabric and includes shirts, pants, vests, and jackets.

horizontal cabling: The cabling that runs from a work area outlet (WAO) to a telecommunications closet.

I

index-matching material: A material in liquid, paste, film, or gel form that has a refractive index that is nearly equal to the core index of a fiber and is used to reduce reflections from a fiber end face.

Information Technology Systems Installation Methods Manual (ITSIMM): A reference and techniques guide published by BICSI.

insertion loss: The loss or degradation of a signal in a VDV cable when a connector is inserted into another connector or associated hardware.

intermediate distribution frame (IDF): A metal rack designed to connect cables and is located in an equipment room or telecommunications room.

International Brotherhood of Electrical Workers (IBEW): A trade organization that represents about 660,000 workers in the electrical, telecommunications, construction, utilities, broadcasting, railroad, manufacturing, and governmental industries.

intersystem bonding termination: A device that provides a means of connecting bonding conductors from communications systems to a grounding electrode system.

J

jacket: The outer protective layer of a fiber-optic cable.

K

Kevlar®: *See* aramid.

knee pad: A rubber, leather, or plastic pad strapped onto the knee to provide protection and comfort as well as reduce fatigue.

L

ladder: A device consisting of two side rails joined at intervals by steps or rungs.

ladder duty rating: A rating that indicates the total weight, or the combined weight of personnel and tools, that a ladder is designed to support under normal use.

laser transmitter: A device that takes an electrical input and converts it to a concentrated light output.

launch cord: A fiber-optic cable used to introduce light from an optical source into another optical fiber.

legend: A description or explanation of symbols and other information on a drawing or print.

light-emitting diode (LED): A semiconductor device that emits light when an electric current is passed through it.

light mode: A physical path taken by a ray of light.

link: An end-to-end transmission path provided by a VDV cabling system.

local area network (LAN): A short distance VDV communications network used to link computers and peripheral devices under a standard control format.

loose-tube cable: A fiber-optic cable in which the buffer tube surrounding the glass strand is larger than the outer coating on the glass.

loss budget: The calculated signal loss that is expected for a fiber-optic installation.

M

macrobend: A bend that occurs when a fiber-optic cable is bent around a large radius.

mandrel: A fiber-wrapping device used to create attenuation within a fiber-optic cable.

MasterFormat®: A master list of numbers and titles for organizing information about construction requirements, products, and activities into a standard sequence.

mechanical splice: A fiber-optic splice in which the ends of two different fibers are mechanically connected.

membrane penetration: A penetration that only penetrates one side of a fire-rated element.

microbend: A small distortion in a fiber-optic cable caused by a crushing or pinching force.

modal dispersion: The slowing of bandwidth caused by the different rates of speed between modes of light through a fiber-optic strand.

multimode fiber: A glass fiber in which the core diameter is large enough to allow multiple light rays to travel through it.

N

National Electrical Code® (NEC®): A nationally accepted guide regarding the safe installation of electrical conductors and equipment. Also known as NFPA 70®.

National Fire Protection Association (NFPA): A national organization that provides guidance on safety and in assessing the hazards of the products of combustion.

near-end crosstalk (NEXT): The measure of the amount of signal interference from one pair into the pair next to it in the same cable.

network interface card (NIC): A printed circuit board comprised of electronic circuitry used to connect a workstation to a local area network (LAN).

NFPA 70®: *See* National Electrical Code® (NEC®).

nonconductive cable: A cable that does not have metal components.

numerical aperture (NA): The light-accepting ability of a glass fiber.

O

110-type cross-connect: A compact connecting device that can be arranged for use with either jumper wires or patch cords.

open pair: A fault that occurs in VDV cabling when one or both conductors of a pair are not connected (open).

optical attenuation test set: A test instrument that is a power meter with cable inputs and is used to measure attenuation, or signal loss, in a fiber-optic cable.

optical loss test set (OLTS): A test instrument that is a combination of a light source and power meter used to detect and measure attenuation in a fiber-optic cable.

optical receiver: A device that converts optical signals back into a replica of the original electrical signal.

optical time domain reflectometer (OTDR): A test instrument used to measure fiber-optic cable attenuation.

optical transmitter: A device that takes an electrical input and converts it to an optical output.

outlet identifier: A unique system of letters and/or numbers used for identifying cables and their locations in a VDV system.

P

parallel optics: A fiber-optic technology design that uses multiple fibers for a single data stream.

patch cord: A flexible 3′ to 12′ length of cable used to connect a network device to a main cable run or a panel.

patch panel: A device used to make connections between incoming and outgoing VDV lines.

pathway: A set of devices used to route telecommunications cables from a telecommunications room or equipment room to a work area.

permanent link: The installed cable, connectors, cross-connects, and outlets in a telecommunications installation project.

personal protective equipment (PPE): Clothing, glasses, gloves, hard hats, respirators, or other safety devices designed to protect workers against safety hazards in the work area and from injury.

photodiode: A semiconductor device that detects light and converts it into an electrical signal.

plan view: A drawing of an object as it appears looking down from a horizontal plane.

plenum: A duct or area connected to distribution ducts through which air moves.

power sum attenuation to crosstalk ratio far-end (PS ACR-F): The ratio of signal strength to undesired signal noise in a twisted-pair cable at the near end of multiple transmit pairs to an adjacent pair, as measured at the far end of the cable.

power sum near-end crosstalk (PS NEXT): A measure of the coupling of undesired signal noise in a twisted-pair cable at the near end of multiple transmit pairs to an adjacent pair, as measured at the near end of the cable.

primary bonding busbar (PBB): A busbar that serves as the central connection point between the grounding electrode system for a building electrical service and the grounding and bonding system for telecommunications.

primary protector: A surge protector that is specifically designed to protect communication cables and the equipment connected to the cables from electrical surges.

propagation delay: *See* delay skew.

protective helmet: A rigid hat made from plastic that is used in the workplace to prevent head injury from impact or penetration by falling and flying objects.

Q

qualification tester: A test instrument designed to determine if a VDV cable has a connectivity problem and if the cable can support the bandwidth requirements of a specific network.

quality assurance (QA): A planned and systematic set of actions necessary to provide adequate confidence that an item or product conforms to established technical requirements.

quality control (QC): A system used to ensure that specified standards, including those for accuracy and quality, are met for manufactured products and materials.

R

receive cord (tail cord): A fiber-optic cable used to measure insertion loss with an OTDR at the far end of a cable plant.

reflectance: An amount of light reflected back along the path of transmission from a connector or terminated fiber. Also known as return loss or back reflection.

refraction: The reflection of light within a glass strand caused by the differing densities of glass used in the core and cladding.

refractive index: The ratio of the velocity of light between the core and the cladding of an optical fiber.

Registered Communications Distribution Designer (RCDD): An individual who has acquired education, training, and expertise in the design, implementation, and integration of telecommunications and data transport systems and infrastructure.

return loss: 1. The ratio of transmitted signal strength to signal strength reflected back to the transmitting end of a channel or permanent link. **2.** *See* reflectance.

ring conductor: The second wire in a pair of wires.

rip cord: A thread installed along a cable jacket or sheath that allows the cable jacket or sheath to be split for access to individual fibers.

riser cabling: *See* backbone cabling.

RJ-45 connector: An 8-pin connector used for data transmission over standard telephone wire.

S

safety barrier: A set of objects such as cones, line markings (delineators) or posts used to mark a potentially hazardous work area.

safety data sheet (SDS): A document that provides detailed information describing a chemical, instructions for its safe use, its potential hazards, proper disposal and appropriate first-aid measures.

safety glasses: An eye-protection device with reinforced frames, side shields, and lenses made of impact-resistant glass or plastic.

secondary bonding busbar (SBB): A busbar located in the telecommunication room or equipment room of a building and connects to the primary bonding busbar.

service entrance facility: The location where telecommunications cabling enters a building.

shielded-pair cable: A cable with color-coded, insulated pairs of wires wrapped in sheaths, which are all wrapped within a metallic braid or foil to prevent the wires from picking up external signals or interference.

shorted pair: A fault that occurs when the two conductors of a VDV cable pair are short-circuited and continue to have electrical continuity.

single-mode fiber: A glass fiber in which the core diameter is sized for only one light ray to travel through it.

sound system: A network of hardwired or wireless speakers installed throughout a building to transmit sound only, such as music, alarm notifications, or paging, throughout a specified location.

specification: Written information that is included with a set of prints.

splice: The joining of two or more fiber-optic strands.

splice tray: A device used to organize and protect spliced fibers.

split pair: A fault that occurs in a VDV cable when the wire in one pair is interchanged with a wire in another pair.

static electricity: An electrical charge at rest.

stepladder: A folding ladder that stands without support. Also known as an A-frame ladder.

strength member: A fiber-optic cable component that runs the length of the cable and is used to increase cable tensile (pulling) strength.

structured cabling: A systematic installation of cabling needed for low-voltage systems.

surge: A type of transient voltage or current that usually rises rapidly to a peak value and then falls more slowly to zero.

surveillance system: A system that controls security cameras.

T

telecommunication: The communication of voice, data, images, and video over a distance via electromagnetic or optical sources on one or more hardwired or wireless networks.

telecommunications: The field of study involving telecommunication.

telecommunications bonding backbone (TBB): A cable that connects a primary bonding busbar to secondary bonding busbars located in telecommunication rooms or equipment rooms located throughout a building.

telecommunications bonding conductor (TBC): A conductor that bonds a primary bonding busbar to the grounding electrode system for a building electrical service.

telecommunications equipment bonding conductor (TEBC): A conductor that connects a primary bonding busbar or secondary bonding busbar to equipment cabinets or racks.

Telecommunications Industry Association (TIA): An international trade group that represents several hundred telecommunications companies.

telecommunications room (TR): An enclosed space that houses telecommunications equipment, cable terminations, and cross-connect cabling used to service work areas on the same floor of a building.

terminator: A device installed on the end of a VDV line that absorbs a signal to prevent it from exiting through an unused port.

through penetration: A penetration that passes completely through a fire-rated element.

tight-buffered cable: A fiber-optic cable that has the glass strand tightly attached to the buffer tube so that they move as a single construct.

tip conductor: The first wire in a pair of wires.

toner: A test instrument designed to identify and locate a copper conductor or cable.

topology: The shape or arrangement of a network system.

transceiver: A device that both transmits and receives.

transient voltage: A temporary, unwanted voltage in an electrical circuit.

Transmission Performance Specifications for Field Testing of UTP Cabling Systems (TSB-67): A document that specifies the electrical characteristics of field test instruments, test methods, and minimum transmission requirements for UTP cabling.

twisted-pair cable: A cable that consists of multiple pairs of insulated copper wires twisted around each other lengthwise to reduce interference from one wire to another.

V

verification tester: 1. *See* wiremap tester. **2.** A test instrument designed to trace the path of conductors and confirm that conductors and data outlets are wired correctly.

visual fault locator: A fiber-optic test tool that illuminates the locations of cable faults with the use of a bright red light source that escapes through faults in the cable jacket, splices, or connectors.

visual light source: A light used to verify of the continuity of a fiber strand.

Voice over Internet Protocol (VoIP): The method and technology used to transmit voice communication over a data network through use of the Internet.

W

wavelength: The distance between two identical points in the adjacent cycles of an oscillating signal.

wire: An individual conductor.

wireless access point (WAP): A device that connects wireless telecommunication devices to form a wireless network.

wiremap (cabling map): A map that shows how VDV cabling is routed throughout a building.

wiremap test: A cable test used to determine the quality of the connection between both ends of a cable.

wiremap tester: A test instrument used to locate open circuits, short circuits, and improper cable terminations such as swapped conductors. Also known as verification tester.

wire spudger: A device used to remove individual wires from a bundle, push wires into position, or remove wires from an insulation displacement contact.

work area: A location where an individual uses a PC, a telephone, and other user equipment devices connected to structured (horizontal) cabling.

Index

Page numbers in italic refer to figures.